Jens Kunstmann

Modeling sp Materials

Jens Kunstmann

Modeling sp Materials

Novel Phases of Elemental Boron, Structure Control of Nanotubes, and the Enatom Method

Südwestdeutscher Verlag für Hochschulschriften

Impressum/Imprint (nur für Deutschland/ only for Germany)
Bibliografische Information der Deutschen Nationalbibliothek: Die Deutsche Nationalbibliothek verzeichnet diese Publikation in der Deutschen Nationalbibliografie; detaillierte bibliografische Daten sind im Internet über http://dnb.d-nb.de abrufbar.
Alle in diesem Buch genannten Marken und Produktnamen unterliegen warenzeichen-, marken- oder patentrechtlichem Schutz bzw. sind Warenzeichen oder eingetragene Warenzeichen der jeweiligen Inhaber. Die Wiedergabe von Marken, Produktnamen, Gebrauchsnamen, Handelsnamen, Warenbezeichnungen u.s.w. in diesem Werk berechtigt auch ohne besondere Kennzeichnung nicht zu der Annahme, dass solche Namen im Sinne der Warenzeichen- und Markenschutzgesetzgebung als frei zu betrachten wären und daher von jedermann benutzt werden dürften.

Verlag: Südwestdeutscher Verlag für Hochschulschriften Aktiengesellschaft & Co. KG
Dudweiler Landstr. 99, 66123 Saarbrücken, Deutschland
Telefon +49 681 37 20 271-1, Telefax +49 681 37 20 271-0, Email: info@svh-verlag.de
Zugl.: Universität Stuttgart, Diss. 2008

Herstellung in Deutschland:
Schaltungsdienst Lange o.H.G., Berlin
Books on Demand GmbH, Norderstedt
Reha GmbH, Saarbrücken
Amazon Distribution GmbH, Leipzig
ISBN: 978-3-8381-1031-8

Imprint (only for USA, GB)
Bibliographic information published by the Deutsche Nationalbibliothek: The Deutsche Nationalbibliothek lists this publication in the Deutsche Nationalbibliografie; detailed bibliographic data are available in the Internet at http://dnb.d-nb.de.
Any brand names and product names mentioned in this book are subject to trademark, brand or patent protection and are trademarks or registered trademarks of their respective holders. The use of brand names, product names, common names, trade names, product descriptions etc. even without a particular marking in this works is in no way to be construed to mean that such names may be regarded as unrestricted in respect of trademark and brand protection legislation and could thus be used by anyone.

Publisher:
Südwestdeutscher Verlag für Hochschulschriften Aktiengesellschaft & Co. KG
Dudweiler Landstr. 99, 66123 Saarbrücken, Germany
Phone +49 681 37 20 271-1, Fax +49 681 37 20 271-0, Email: info@svh-verlag.de

Copyright © 2009 by the author and Südwestdeutscher Verlag für Hochschulschriften Aktiengesellschaft & Co. KG and licensors
All rights reserved. Saarbrücken 2009

Printed in the U.S.A.
Printed in the U.K. by (see last page)
ISBN: 978-3-8381-1031-8

Für meine Familie

4

Contents

Abbreviations		9
1 Introduction		**11**
2 Theoretical Methods		**17**
2.1	Introduction	17
2.2	Hamiltonian	18
2.3	Approximations	18
	2.3.1 Adiabatic Approximation	18
	2.3.2 Density Functional Theory	22
	2.3.3 Density Functional Perturbation Theory	27
2.4	Periodic Solids	29
	2.4.1 Energy Band Model	30
	2.4.2 Lattice Vibrations	32
	2.4.3 Electron–Phonon Coupling	34
	2.4.4 Superconductivity	36
2.5	Basis Sets	37
	2.5.1 Linear Muffin Tin Orbitals	37
	2.5.2 Plain Waves and Pseudopotentials	42
	2.5.3 Discussion	47
3 Novel Phases of Elemental Boron		**49**
3.1	Introduction	49
3.2	Fundamentals and Methods	50
	3.2.1 Boron Chemistry	50
	3.2.2 Elemental Bulk Phases	53
	3.2.3 Clusters, Nanostructures, and the Aufbau Principle	58
	3.2.4 Computational Details	61
3.3	Broad Sheets	64
	3.3.1 Finding Structure Models	65
	3.3.2 The Planar Boron Sheet	68

		3.3.3	The Puckered Boron Sheet	69

        3.3.3 The Puckered Boron Sheet 69
        3.3.4 Summary 73
  3.4 Nanotubes 74
        3.4.1 Ideal Boron Nanotubes 74
        3.4.2 Real Boron Nanotubes 80
        3.4.3 Strain Energy 88
        3.4.4 Summary 90
  3.5 Layered Bulk Phases 92
        3.5.1 Introduction: High Pressure and Superconductivity 92
        3.5.2 Crystal Structures and Chemical Bonding 94
        3.5.3 Phase Diagram at $T=0$ K 99
        3.5.4 Electronic Structure, Phononic Structure, and Superconductivity 103
        3.5.5 Summary 121
  3.6 Summary and Conclusions 122
  3.7 Outlook 126

4 Structure Control of Nanotubes 129
  4.1 Introduction 129
  4.2 Strain Energy 130
        4.2.1 Carbon Nanotubes 130
        4.2.2 Boron Nanotubes 132
  4.3 Structure Control 134
  4.4 Summary 134

5 The Enatom Method 137
  5.1 Introduction 137
        5.1.1 Theoretical Background 138
        5.1.2 Physical Motivation 140
  5.2 Computational Details 142
  5.3 Results and Discussion 143
        5.3.1 Rigid and Deformation Part 144
        5.3.2 Rigid Part 147
        5.3.3 Deformation part 152
  5.4 Summary and Outlook 156

6 Summary and Outlook 159

A Enatom Quantities 165

Bibliography 183

Acknowledgments 185

Abbreviations

ASA	atomic sphere approximation
bct	body centered tetragonal
bcc	body centered cubic
BNT	boron nanotube
BS	boron sheet
BZ	Brillouin zone
CNT	carbon nanotube
DFT	density functional theory
DFPT	density functional perturbation theory
DOS	density of states
fcc	face centered cubic
FS	Fermi surface
GGA	generalized gradient approximation
hcp	hexagonal closed packed
LDA	local density approximation
LMTO	linear muffin tin orbital
MT	muffin tin
MTO	muffin tin orbital
LR	linear response
PT	perturbation theory
SMTO	single muffin tin orbital
TB	tight–binding
T_c	critical temperature of a superconductor

Chapter 1
Introduction

Materials whose valence states consist of s and p electrons only are commonly referred to as sp materials. Those materials have recently attracted considerable interest in different fields science. For instance, the discovery of carbon fullerenes [1] and carbon nanotubes [2], and the existence of stable multilayers or single layers of graphene [3] have lead to the development new research fields in physics and chemistry. Another example is the discovery of $T_c = 39$ K superconductivity in the the simple binary sp compound MgB_2 [4], which was a big surprise for the scientific community and raised huge interest worldwide. Lithium under high pressure is a further example. At ambient conditions it is a simple free–electron–like sp metal and a normal conductor. Under pressure, however, it undergoes several interesting phase transitions [5], and becomes a superconductor with a pressure dependent T_c reaching up to 20 K at about 50 GPa [6, 7, 8]. Also elemental boron, a semiconductor at ambient conditions, transforms to a superconductor under pressure with T_c reaching 11 K at 250 GPa [9].

In this thesis we present results that are relevant to nanoscience, materials under high–pressure, superconductivity, and chemistry. We will mainly focus on boron and study its properties in a general framework. This is exceedingly necessary because elemental boron is little studied and many fundamental properties such as the phase diagram, the ground state structure, or its high–pressure behavior are unknown [10, 11, 12, 13]. Nevertheless, boron is a very fascinating element as it forms bulk structures of remarkably complexity (three–dimensional networks of B_{12} icosahedra) and it has a chemical versatility which is unique among the elements of the periodic table [14]. This is primarily due to its electron deficient nature [15], i.e., it has four valence orbitals (one s and three p orbitals) but only three valence electrons. The usual two–center σ bonds are inefficient to solve the problem of electron deficiency. Therefore elemental and boron–rich compounds are primarily held together by multi–center bonds [16, 17, 18], where three or more atomic orbitals belonging to different atoms combine to give a single lowest–energy state and bond charge is accumulated around the center of gravity of all atoms involved. The actual bonding in boron materials is a complex interplay of σ, π, and the dominating multi–center bonds [18]. However, a proper understanding of the chemical bonding in its bulk structures going beyond these basic ideas is still lacking.

A surprising development in boron chemistry comes from studies of elemental clusters, that were found to form sheet–like, quasiplanar structures [19, 20]. In addition, the first successful synthesis of boron nanotubes was reported [21]. These discoveries were anticipated by theoretical studies of Boustani and Quandt in the mid–1990s that culminated in the formulation of a so called Aufbau principle by Boustani [22]. It is a structural rule that predicts the existence of quasiplanar (sheets) [23], tubular (nanotubes) [24, 25], convex and spherical (fullerenes) [26] boron clusters. The Aufbau principle shifted the conventional paradigm that boron structures are based on B_{12} icosahedra. However, it is still a very general approach and the corresponding experimental studies are not very detailed yet. Therefore questions about the precise atomic structure of boron nanotubes and boron sheets remain open, and further theories describing their properties are needed.

In this thesis we will try to arrive at a better understanding of these novel nanostructures of boron. We start with the following simple consideration: If we know that small boron clusters are quasiplanar, we can immediately conclude that boron nanotubes and boron fullerenes should exist, because a (quasi)planar cluster that is growing in size tends to remove dangling bonds by forming closed tubular or polyhedral modifications. This is completely analogous to carbon, where fullerenes and nanotubes only grow out of the cluster phase, and are a compromise between the energy cost for bending the planar carbon cluster, and the energy gain for removing dangling bonds. As graphene, a broad carbon sheet, is the precursor of carbon nanotubes, a broad boron sheet will be the precursor of boron nanotubes. So the missing link between boron nanotubes and the quasiplanar clusters is a broad boron sheet, the limiting case of a quasiplanar clusters for an infinite number of atoms. We will determine the structure and the properties of that broad boron sheet and apply our findings to the related boron nanotubes in order to predict their basic properties.

The existence of boron nanotubes and the prediction of boron fullerenes might eventually lead to a new field of research based on boron nanostructures. In some sense nanoscience (nanotechnology) is a trend opposite to miniaturization. In the former one tries to build up complex macrostructures and functional units from small basic elements that are atoms or molecules (bottom-up approach), in the latter one tries to reduce the size of existing technologies to smaller and smaller sizes (top-down approach).[1] The miniaturization of electronic circuits for example has lead to an

---

[1]It is interesting to mention that many ideas of nanoscience go back to Richard Feynman. At the annual meeting of the American Physical Society on December 29 in 1959, he gave a speech with the title "There's Plenty of Room at the Bottom" [27]. With the subtitle "An Invitation to Enter

exponential increase in computer power since the 1960s, and the number of transistors that can be inexpensively placed on an integrated circuit is doubling approximately every two years (Moore's "law" [28]). But this trend will come to an end if the size of individual transistors (at present around 50 nm) becomes comparable to the atomic size. For future technologies the bottom-up approach of nanoscience obviously has a large potential. However, its biggest problem is the lack of direct control over the atomic structure of materials. An example that illustrates the problem of structure control are carbon nanotubes. Their electronic and mechanical properties depend strongly on their atomic structure, which is characterized by two parameters: the radius and the chiral angle (chirality) [29]. For the standard synthesis of carbon nanotubes, one may achieve some control over their radii, but little control over their chiralities [30, 31, 32], which implies that in general, there is little control over the properties of the end products of the synthesis. As carbon nanotubes are either metallic or semiconducting, depending on their radii and chiralities [29], this poor structure control also implies a poor control over their electronic properties. In practice, metallic and semiconducting nanotubes are separated by cumbersome trial and error procedures. Such approaches can hardly be regarded as a "technology" and there is a strong need for better control over the atomic structure of materials. Our ability to achieve structure control will decide whether nanoscience will eventually lead to a useful technology. On the atomic scale, however, conventional industrial techniques are not applicable and entirely new technologies are required. For that purpose one probably has to rely on chemical or biological self-organization processes and use specific materials that allow for such processes to actually take place. Here it would be desirable to link macroscopic quantities that we can control (e.g. the chemical reaction conditions) to microscopic structural parameters of these materials (e.g. the radius and chirality of nanotubes). We will illustrate this idea in more detail in this thesis.

Carbon nanotubes and carbon fullerenes can be related to graphene, which is a single sheet of graphite. If we consider a broad boron sheet to be the boron analog of graphene (with respect to their laminar structures), and face that boron nanotubes exist [21] (and that boron fullerenes were predicted [26, 33, 34]), it is likely that layered bulk phases similar to graphite may also exist for boron. Such structures would be quite different from the known bulk phases, that consist of complex three-dimensional networks of B_{12} icosahedra. However, the pronounced polymorphism of elemental boron [35] makes it probable that novel, so far undiscovered phases exist. If these structures are not stable at ambient conditions, they might be so at high pressures. We will therefore extend the ideas behind the Boustani Aufbau principle, which was only developed for elemental clusters, to the bulk domain, and speculate

---

a New Field of Physics" he asked "Why cannot we write the entire 24 volumes of the Encyclopedia Brittanica on the head of a pin?"

whether layered bulk phases of boron, a new family of boron bulk materials, may exist. This is done by constructing layered phases from the broad boron sheet and by a literature search for boron structures that fall into the class of layered bulk materials.

In the study of bulk structures, we again note that elemental boron, a semiconductor at ambient conditions, transforms to a superconductor under pressure with T_c reaching 11 K at 250 GPa [9]. This is interesting with respect to the superconductivity in MgB$_2$ (T_c = 39 K) [4] and in boron doped diamond (T_c = 4 K) [36]. Although the mechanism driving the superconductivity in boron doped diamond can be considered as a three dimensional variant of the one in MgB$_2$ [37, 38], the role boron plays in the two compounds is quite different and it is not very likely that the superconductivity in boron under high pressure is based on a similar mechanism. The biggest problem for explaining the superconductivity in boron is the general lack of knowledge about its high–pressure phases, and the corresponding crystal structures are still under debate. Thus the problem of superconductivity merges with that of studying bulk structures. So far, three different theoretical approaches were used to determine possible high–pressure phases. One is based on studying the high–pressure behavior of the common icosahedral bulk structures [39, 40], another on randomly trying different naive phases such as fcc, bcc, etc. [41, 42, 43], and a third approach assumes that boron under pressure adopts similar structures than heavier group-III elements (Al, Ga, In) [44, 45, 46]. We will try a new approach and study the high–pressure behavior of the layered bulk materials, introduced above. The latter are constructed from our basic understanding of boron chemistry. We will test these new phases by determining their thermodynamic stability in comparison with other bulk phases, and by calculating phonon dispersions to determine their dynamical stability. Furthermore, we will study the electron–phonon coupling in the layered structures and see if those effects are responsible for the high–pressure superconductivity of elemental boron.

The electron–phonon coupling, i.e., the strength of the interactions between electrons and phonons, and the energy dispersion of phonons are commonly calculated via linear response methods [47, 48, 49]. They provide an efficient way to determine physical properties of solids that are related to lattices dynamics. However, linear response methods are usually based on an abstract reciprocal space formulation. The problem of studying superconductivity in elemental boron under pressure shows that often a reciprocal space picture of the coupling is not enough. We would like to have a method that gives a real–space understanding of of a solid, its vibrational properties, and the electron–phonon interactions. Such a method was proposed in the 1970s by Ball [50, 51], but never realized in practice. Ball's method allows to describe condensed matter as a collection of generalized (pseudo)atoms which are constructed from the linear response to atomic displacements from equilibrium positions. Such a pseudoatom, which we will call *enatom*, consists of a rigid and a deformation density (and potential). The rigid part defines a unique decomposition of the equilibrium density (or potential) into atomic–like but overlapping contributions that move rigidly, with the nuclear position. The deformation density (potential) describes how this charge (potential) deforms upon a displacement, and can be viewed as a backflow. Here, we will provide the first explicit examples of the enatom density and potential, and study their pressure evolution in lithium and aluminum. At ambient pressure both elements are simple free–electron–like sp metals; Li is bcc and a normal conductor [52], Al is fcc and a superconductor with $T_c = 1.2$ K [53]. Under pressure, however, the two systems evolve very differently. The electronic structure of Al remains that of a free electron-like metal, its superconductivity is suppressed (with $T_c < 0.1$ K at 6 GPa [53]), and a structural transition to a hcp phase takes place only at $P > 217$ GPa [54]. Li, on the other hand, becomes increasingly covalent,[2] it undergoes several phase transitions [5], and becomes a superconductor with a pressure dependent T_c reaching up to 20 K at about 50 GPa [6, 7, 8]. Linear response studies have established that such a T_c results from a large increase of electron–phonon coupling under pressure [55, 56, 57, 58]. We therefore expect that the enatom of the two systems will display different behaviors under pressure. Lithium and aluminum thus provide a simple and still interesting starting point for the study of enatom quantities. Further applications of the enatom method to boron systems should follow in the future.

All our studies employ density functional theory (DFT) [59, 60] within the framework of the local density approximation (LDA) [60, 61, 62] or the generalized gradient approximation (GGA) [63, 64, 65]. Experience has shown that these approximations describe ground state properties (e.g. total energies, charge densities, etc.) quite well, but they do not correctly describe, for example, orbital energies, band gaps in semiconductors and insulators, excited states, localized d or f electrons, strong electronic correlations, or long–ranged interactions as van der Waals forces or interactions via Hydrogen bonds [66, 67]. However, all of those failures are unimportant for the questions we would like to answer, as the rather delocalized sp electrons are well described by the LDA or the GGA. DFT provides a parameter–free quantum mechanical (first principles) description of materials that allows us to determine many material properties. This will enable us to study the atomic structure, the band structure, the chemical bonding, the phase diagram, and mechanical and thermodynamic properties of the sp materials in consideration. Phonon energy dispersions and electron–phonon couplings are calculated via density functional perturbation theory (linear response) because this approach is a lot more efficient than a direct (frozen–phonon) DFT calculation [47]. All theoretical methods that are used throughout this thesis will be described in chapter 2.

---

[2]In the case of metals we use the term "covalency" in a loose sense to indicate the appearance of directional bonds.

In chapter 3 we study broad sheets, nanotubes, and layered bulk phases of elemental boron. After a general introduction to our present knowledge about boron, we use structural optimization methods to establish a simple model for a broad and stable boron sheet. We then analyze its properties and show how these results may be used to predict the structure, stability, electronic and mechanical properties of boron nanotubes. Our findings will define a consistent picture of boron sheets and boron nanotubes, which unifies former studies on these materials in the framework of a generalized theory. Then the high–pressure behavior, the structure, the stability, and the electron–phonon coupling of three layered bulk phases are studied. We show that such structures are likely to exist at elevated pressures or even at ambient conditions, and that there are strong indications that they could be conventional superconductors. Furthermore, by analyzing the similarities of the chemical bonding in the common icosahedral and the new layered phases of boron, we were able to define a generalized picture of the chemical bonding in elemental boron solids.

In chapter 4 we are concerned with the structure control of nanotubular materials. We explain why one has very limited control over the structure of carbon nanotubes during their synthesis. Then we show that novel classes of nanotubes, which are related to sheets with anisotropic in–plane mechanical properties (e.g. boron nanotubes), could overcome those problems. Our results further suggest that extended searches for nanotubular materials similar to pure boron might allow for one of the simplest and most direct ways to achieve structure control.

In chapter 5 we provide the first explicit examples of the enatom density and potential for fcc lithium and fcc aluminum, at pressures of 0, 35, and 50 GPa. We analyze the relative importance of the rigid and deformation parts of the density and the potential, and determine the degree of sphericity of the rigid parts. We further show that the basic features of the spherical part of the rigid density and potential can be understood by means of linear screening theory, and we find that the lattice symmetry determines the structure of the deformation parts of an enatom in fcc Al and fcc Li.

Finally, a summary of all our results and an outlook will be given in chapter 6.

Chapter 2

Theoretical Methods

In this chapter we want to familiarize the reader with the theoretical methods that underlie this thesis. This includes the quantum mechanical *ab initio* methods density functional theory and density functional perturbation theory, the theoretical description of periodic solids, the used basis sets, and the physical quantities that are discussed in the subsequent chapters.

2.1 Introduction

The term *ab initio* means *from first principles*. Does that mean that the methods dealt with in this chapter are able to solve the fundamental equations of quantum mechanics in their most general from? Not really, as this would exceed the available computing power of current computers significantly. Even *ab initio* methods use approximate variants of quantum mechanics and many simplifications. However, they are used in such a comprehensive way that the theory is free of undetermined parameters, except the fundamental physical constants. Therefore, the term *ab initio* should rather be understood as *without free parameters*.

2.2 Hamiltonian

The main object of interest is the following fundamental Hamiltonian for a system of N electrons and Γ nuclei which is able to describe atoms, molecules, and solids.

$$\begin{aligned}\hat{H} = \hat{H}(r,R) &= \hat{T}_E + \hat{T}_N + \hat{W}_{EE} + \hat{W}_{EN} + \hat{W}_{NN} \\ &= -\frac{1}{2}\sum_{i=1}^{N}\nabla_i^2 - \frac{1}{2}\sum_{\alpha=1}^{\Gamma}\frac{\nabla_\alpha^2}{M_\alpha} + \sum_{i<j}^{N}\frac{1}{|\mathbf{r}_i - \mathbf{r}_j|} - \sum_{i=1}^{N}\sum_{\alpha=1}^{\Gamma}\frac{Z_\alpha}{|\mathbf{r}_i - \mathbf{R}_\alpha|} \\ &\quad + \sum_{\alpha<\beta}^{\Gamma}\frac{Z_\alpha Z_\beta}{|\mathbf{R}_\alpha - \mathbf{R}_\beta|}\end{aligned} \quad (2.1)$$

Here $r = \{\mathbf{r}_i\}$ represents the set of all electronic coordinates (Latin indices) and $R = \{\mathbf{R}_\alpha\}$ the set of all nuclear coordinates (Greek indices). The nabla operators ∇_i and ∇_α act on the particle positions $\mathbf{r}_i$ and $\mathbf{R}_\alpha$, respectively. M_α is the mass and Z_α the atomic number of the nucleus with index α. Now and in the following we will use atomic units where $\hbar = e = m_e = 1$.

The different terms in Eq. 2.1 have the following physical meaning: $\hat{T}_E$ is the kinetic energy of the electrons, $\hat{T}_N$ is the kinetic energy of the nuclei, $\hat{W}_{EE}$ and $\hat{W}_{NN}$ represent the electrostatic Coulomb repulsion among electrons and nuclei, respectively and $\hat{W}_{EN}$ stands for the attractive Coulomb interactions between the negatively charged electrons and the positively charged nuclei.

As Hamiltonian 2.1 is explicitly time–independent it will be treated with time–independent quantum theory in the following.

2.3 Approximations

As already mentioned in the introduction, the quantum mechanical system defined by the Hamiltonian 2.1 is much too complicated to be solved directly. Therefore we will now describe the approximations that are used to simplify the theoretical treatment.

2.3.1 Adiabatic Approximation

Fundamentals

A fundamental approximation in molecular and solid state physics is the adiabatic approximation which is also called the *Born-Oppenheimer approximation*. It is based in the physically intuitive picture that electrons and nuclei move on different time scales due to their big difference in mass. One assumes that the fast electronic degrees of freedom instantaneously follow the movements of the nuclei. In other words, the

2.3. APPROXIMATIONS

relaxation time of the electronic subsystem is considered to be zero and the electrons are always in their so called *instantaneous ground state*. For the electrons, on the other hand, the nuclear positions are fixed.

Based on the small parameter m_e/M_α it is possible to set up a perturbation series where in first order the degrees of freedom of electrons and nuclei are decoupled [68]. The wave function is separated into an electronic part $\Psi_E(r,R)$ and nuclear part $\Psi_N(R)$

$$\Psi(r,R) = \Psi_E(r,R)\Psi_N(R). \tag{2.2}$$

The electronic Hamiltonian is

$$\hat{H}_E(r,R) = \hat{T}_E(r) + \hat{W}_{EE}(r) + \hat{W}_{EN}(r,R) + \hat{W}_{NN}(R). \tag{2.3}$$

It follows from eliminating $\hat{T}_N$ in $\hat{H}$. The corresponding Schrödinger equation reads

$$\hat{H}_E(r,R)\Psi_E(r,R) = E(R)\Psi_E(r,R). \tag{2.4}$$

Since the nuclei are rigid for the electrons the nuclear positions R in Eq. 2.4 are fixed parameters. Now, the nuclear Hamiltonian is

$$\hat{H}_N(R) = \hat{T}_N(R) + E(R). \tag{2.5}$$

Here the energy of the electronic system $E(R)$ is the potential for the nuclei. Therefore it is also called the *interatomic potential* (alternative names are *Born-Oppenheimer energy surface* or *energy landscape*). This leads to the descriptive picture of the nuclei moving in a "glue" of electrons. The latter mediates attractive forces between the former and thus the chemical bonding. The Schrödinger equation for the nuclei finally is

$$\hat{H}_N(R)\Psi_N(R) = \epsilon\,\Psi_N(R). \tag{2.6}$$

Ionic Subsystem

Atomic Forces For the calculation of atomic forces the quantum mechanical character of the nuclear Hamiltonian 2.5 is usually neglected and it is treated classically. Thus the nuclei are considered to be classical point particles moving in the potential $E(R)$ and the forces on them are

$$\mathbf{F}_\alpha = -\nabla_\alpha\,E(R_n). \tag{2.7}$$

Here R_n is a certain rigid configuration of the nuclei. In the *Hellmann–Feynman theorem* [69, 70], this expression is evaluated as

$$\begin{aligned}\mathbf{F}_\alpha^{HF} &= -\left\langle\Psi_E\left|\nabla_\alpha\hat{H}_E\right|\Psi_E\right\rangle \\ &= -\int\nabla_\alpha v(\mathbf{r})\rho(\mathbf{r})d\mathbf{r} - \nabla_\alpha\hat{W}_{NN},\end{aligned} \tag{2.8}$$

where all nuclei are in the configuration R_n. Here $v(\mathbf{r})$ is the external potential (defined below in Eq. 2.17), $\rho(\mathbf{r})$ is the electronic ground state charge density (defined in Eq. 2.19), and the gradients of $v(\mathbf{r})$ and $\hat{W}_{NN}$ are simple analytical expressions. Thus to obtain Hellmann–Feynman forces, only the knowledge of the ground state charge density is necessary.

The Hellmann-Feynman theorem requires the basis set, representing the electronic wave functions, to be complete. Incompleteness can lead to significant errors in the forces [71]. These problems can be circumvented by using for example the *Andersen force theorem* [72, 73] which is relatively insensitive to imprecisions in the wave functions.

Structural Optimization If the forces acting on the atoms (nuclear positions) are known, it is possible to move them along their Newtonian trajectories until all forces are zero such that the atoms occupy their equilibrium positions. This would be the simplest kind of a *structural optimization*.

In a structural optimization one starts from a "guessed" initial configuration of nuclear positions R_1 of a certain system and tries to find its isomers R_I where all atomic forces are zero. The isomers are local minima of the potential $E(R)$, i.e., minimizers of the system's total energy. They are defined as

$$\mathbf{F}_\alpha = -\nabla_\alpha E(R_I) = 0, \quad \alpha = 1\ldots\Gamma. \tag{2.9}$$

So, formally the problem is to find local minima of the potential $E(R)$. This is a standard problem of optimization and a number of algorithms are known such as the method of deepest descents or the conjugate–gradients technique [74]. If the initial configuration R_1 is close to a structural optimum, these methods will reliably find it. However, if R_1 is far from an optimum, none of them can guaranty to find the "right" global minimum or just an unimportant local minimum. In the latter case the result will strongly depend on the initial configuration R_1 and the method in use.

Vibrational Properties To study vibrations of molecules and solids the interatomic potential $E(R)$ is expanded in a Taylor series about the nuclear equilibrium positions R_I. This is a reasonable approximation in many cases, since the nuclear displacements are small compared to the interatomic distances. For that purpose we define the position of nucleus α as $\mathbf{R}_\alpha = (R_{\alpha x}, R_{\alpha y}, R_{\alpha z}) = \mathbf{R}_\alpha^I + \mathbf{u}_\alpha$, where $\mathbf{u}_\alpha$ is the Cartesian displacement of nucleus α from its equilibrium position $\mathbf{R}_\alpha^I$. With $u = \{\mathbf{u}_\alpha\}$ a second-order Taylor expansion reads

$$\begin{aligned} E(R_I + u) &= E(R_I) + \sum_{\alpha=1}^{\Gamma} \mathbf{u}_\alpha \cdot \nabla_\alpha E(R_I) + \frac{1}{2} \sum_{\alpha,\beta=1}^{\Gamma} \sum_{\mu,\nu=x,y,z} \frac{\partial^2 E(R_I)}{\partial R_{\alpha\mu} \partial R_{\beta\nu}} u_{\alpha\mu} u_{\beta\nu} \\ &= E(R_I) + 0 + \hat{W}_{\text{harm}} \end{aligned} \tag{2.10}$$

2.3. APPROXIMATIONS

The constant zeroth order term is the total energy of the undisplaced system and defines the zero point of our energy scale. As the nuclei are assumed to be in their equilibrium positions the first order term vanishes due to Eq. 2.9. The second order term $\hat{W}_{\text{harm}}$ is called the harmonic term and since higher order terms are neglected here, the approximation is called the *harmonic approximation*. If the pair $(\alpha\mu)$ is considered as single index the second order coefficients form a $3\Gamma \times 3\Gamma$ matrix which is called the matrix of *interatomic force constants* or *Hesse matrix/Hessian*:

$$H_{\alpha\mu,\beta\nu} = \frac{\partial^2 E(R_I)}{\partial R_{\alpha\mu}\partial R_{\beta\nu}} = -\frac{\partial F_{\alpha\mu}}{\partial R_{\beta\nu}} \quad (2.11)$$

$$H_{\alpha\mu,\beta\nu}^{\text{HF}} = \int \frac{\partial^2 v(\mathbf{r})}{\partial R_{\alpha\mu}\partial R_{\beta\nu}} \rho(\mathbf{r})d\mathbf{r} + \int \frac{\partial v(\mathbf{r})}{\partial R_{\alpha\mu}} \frac{\partial \rho(\mathbf{r})}{\partial R_{\beta\nu}} d\mathbf{r} + \frac{\partial^2 \hat{W}_{NN}}{\partial R_{\alpha\mu}\partial R_{\beta\nu}} \quad (2.12)$$

$H_{\alpha\mu,\beta\nu}$ can be obtained in different ways. One is to transform the derivatives in definition 2.11 into finite differences and then to calculate the interatomic potential $E(R)$ or the atomic forces $F_{\alpha\mu}$ for *all* infinitesimally displaced atomic configurations $R_{\beta\nu} = R_{\beta\nu}^I + u_{\beta\nu}$. For periodic solids this approach is called a *frozen phonon* calculation (see Sec. 2.4.2). A different approach is given by Eq. 2.12, that is obtained by differentiating the Hellmann–Feynman forces (Eq. 2.8) with respect to the nuclear coordinates [47]. The derivatives of the external potential $v(\mathbf{r})$ and $\hat{W}_{NN}$ are again simple analytical expressions. Thus obtaining the Hesse matrix $H_{\alpha\mu,\beta\nu}^{\text{HF}}$ via the Hellmann–Feynman theorem requires the knowledge of the ground state charge density $\rho(\mathbf{r})$ as well as its first derivative $\partial\rho(\mathbf{r})/\partial R_{\beta\nu}$ with respect to an atomic displacement. Equation 2.12 is not valid in general because the basis set cannot always considered to be complete. Especially for atom–centered basis sets (see Sec. 2.5.1) correction terms due to the change of the basis set upon the atomic displacement have to be added to Eq. 2.12.

It is now convenient to introduce mass–weighted displacements $\mathbf{w}_\alpha = (w_{\alpha x}, w_{\alpha y}, w_{\alpha z}) = \sqrt{M_\alpha}\mathbf{u}_\alpha$, where M_α is the mass of nucleus α. If the $\mathbf{w}_\alpha$ are the dynamical variables, the kinetic energy operator for the nuclei and the harmonic potential become

$$\hat{T}_N = -\frac{1}{2}\sum_{\alpha=1}^{\Gamma}\sum_{\mu=x,y,z}\left(\frac{\partial}{\partial w_{\alpha\mu}}\right)^2 \quad (2.13)$$

$$\hat{W}_{\text{harm}} = \frac{1}{2}\sum_{\alpha,\beta=1}^{\Gamma}\sum_{\mu,\nu=x,y,z} H'_{\alpha\mu,\beta\nu}\, w_{\alpha\mu}w_{\beta\nu}$$

$$\text{with:} \quad H'_{\alpha\mu,\beta\nu} = \frac{1}{\sqrt{M_\alpha M_\beta}} H_{\alpha\mu,\beta\nu} \quad (2.14)$$

Using the above expansion in the nuclear Hamiltonian 2.5 and treating it classically yields a system of 3Γ coupled linear second–order equations of motion. They can be decoupled with a vibrational ansatz for the (mass–weighted) atomic displacements

$$w_{\alpha\mu} = \epsilon_{\alpha\mu}\, e^{i\omega t}, \tag{2.15}$$

where the $\epsilon_{\alpha\mu}$ are called *polarizations*. This finally leads to the eigenvalue problem

$$H'\epsilon_j = \omega_j^2 \epsilon_j. \tag{2.16}$$

Solving it yields $j = 1\ldots 3\Gamma$ eigenvalues ω_j^2 and eigenvectors ϵ_j (polarization vectors[1]) and ω_j is the frequency of the vibrational eigenmode j [75, 68].

Electronic Subsystem

Now and in the following paragraphs we will consider the electronic subsystem, defined by Eqs. 2.3 and 2.4, where the nuclear positions R are merely fixed parameters. Therefore we will simplify our notation and no longer specify R explicitly. The interatomic potential $E(R)$ will be called the *total energy* E and the constant term $\hat{W}_{NN}$ in Hamiltonian 2.3 will be omitted for convenience. However, it should be kept in mind that $\hat{W}_{NN}$ is an important contribution to the total energy.

Now the physical picture of the electronic Hamiltonian 2.3 is that of N interacting electrons that move in a *fixed* electrostatic potential generated by Γ nuclei. This potential can therefore be considered as an *external potential*

$$v(\mathbf{r}) = -\sum_{\alpha=1}^{\Gamma} \frac{Z_\alpha}{|\mathbf{r} - \mathbf{R}_\alpha|} \tag{2.17}$$

and Eq. 2.3 can be written in its usual form as

$$\begin{aligned}\hat{H}_E &= \hat{T}_E + \hat{U}_E + \hat{W}_{EE} \\ &= -\frac{1}{2}\sum_{i=1}^{N}\nabla_i^2 + \sum_{i=1}^{N} v(\mathbf{r}_i) + \sum_{i<j}^{N} \frac{1}{|\mathbf{r}_i - \mathbf{r}_j|}\end{aligned} \tag{2.18}$$

where $\hat{U}_E$ is the potential operator and Eq. 2.4 simply becomes $\hat{H}_E \Psi_E = E\Psi_E$. In the following we will only consider ground state properties and E and Ψ_E only refer to the ground state of the system.

2.3.2 Density Functional Theory

The origin of density functional theory (DFT) goes back to Thomas and Fermi [76, 77] and their idea that the electron density

$$\rho(\mathbf{r}) = N\sum_s \int \Psi_E(\mathbf{r}, s; x_2; \ldots; x_N)\Psi_E^*(\mathbf{r}, s; x_2; \ldots; x_N) dx_2 \ldots dx_N \tag{2.19}$$

---
[1] The polarization vectors ϵ_j are 3Γ-dimensional vectors.

2.3. APPROXIMATIONS

contains the essential information about the electronic system in the ground state. The variable $x_i = \mathbf{r}_i, s_i$ represents the spatial $\mathbf{r}_i$ and spin s_i degrees of freedom of particle i, and $\int dx = \sum_s \int d\mathbf{r}$. Equation 2.19 shows the big advantage of that approach: The ground state of the system is not described by a complicated function Ψ_E that depends on $3N$ spatial degrees of freedom and the spin components, but by a function ρ that only depends on the three components of the vector $\mathbf{r}$.

Hohenberg-Kohn Theorem

What started off as a conjecture by Thomas and Fermi could be proven later by Hohenberg and Kohn [59]. Their theorem states that for every given ρ there is at most one external potential v for which ρ is the ground state density. In short: *The external potential $v(\mathbf{r})$ is a unique function of the ground state density $\rho(\mathbf{r})$; $v = v[\rho]$.*[2] The theorem only holds for densities which are "v–representable". These are densities that come from antisymmetric N–electron wave functions that are ground states of Hamiltonian 2.18 with an external potential v (for proofs see [59, 78, 79, 80]); this implies that the potential v must be able to bind N electrons.

This formulation of the Hohenberg–Kohn theorem allows the ground state to be degenerate. So one external potential can lead to different ground states with different electron densities. In contrast, different external potentials can never lead to the same density. If a degeneracy of the ground state is excluded, the mapping between v and ρ is one–to–one (except for a constant) and every external potential can be associated with a single density.

To illustrate this theorem let us consider that Hamiltonian 2.18 is uniquely defined by the number of electrons N and the external potential $v(\mathbf{r})$. Since, according to the Hohenberg–Kohn theorem, the potential is a function of the density and N follows from $N = \int \rho(\mathbf{r})d\mathbf{r}$ the electron density is sufficient to fully determine the Hamiltonian of a system.

Even if the ground state is degenerate the energy E is the same for the different ground states Ψ_E. Therefore it can be expressed as functional of ρ

$$\begin{aligned} E = \langle \Psi_E | \hat{H}_E | \Psi_E \rangle = E[\rho] &= U[\rho] + T[\rho] + W[\rho] \\ &= U[\rho] + F_{HK}[\rho], \end{aligned} \quad (2.20)$$

where the functionals U, T, and W represent the external potential, the kinetic energy, and the electron–electron interactions, respectively. $F_{HK}[\rho] = T[\rho] + W[\rho]$ is the *Hohenberg–Kohn density functional* which is only defined for "v–representable" densities (see above). It is a universal functional that only depends on the number of electrons N. Unfortunately, its is not known explicitly. The external potential is

---
[2] Potentials that differ by a trivial constant are assumed to be equal here.

contained in $U[\rho]$, whose general form is

$$U[\rho] = \langle \Psi_E | \hat{U}_E | \Psi_E \rangle = \int v(\mathbf{r})\rho(\mathbf{r})d\mathbf{r}. \qquad (2.21)$$

Kohn–Sham Method

Since the functional F_{HK} is unknown DFT could not be used in practical calculations in the very beginning. This situation was changed by Kohn and Sham [60]. They introduced a non–interacting auxiliary system described by the Hamiltonian

$$\hat{H}_E^0 = -\frac{1}{2}\sum_{i=1}^N \nabla_i^2 + \sum_{i=1}^N v(\mathbf{r}_i). \qquad (2.22)$$

It is well known from general quantum mechanics how to deal with non–interacting systems: The antisymmetric wave function $\Psi_E^0(x_1,\ldots,x_N)$ can be expressed as Slater determinant $\Psi_E^0 = \frac{1}{\sqrt{N!}} \det[\phi_1 \ldots \phi_N]$ of single particle wave functions $\phi_1(\mathbf{r}_1),\ldots,\phi_N(\mathbf{r}_N)$ called *orbitals*. Inserting the Slater determinant to Eq. 2.19 yields for the electron density

$$\rho(\mathbf{r}) = \sum_i^{occ} |\phi_i(\mathbf{r})|^2, \qquad (2.23)$$

where the index i represents a complete set of one–particle quantum numbers (including spin) and the summation is over the *occupied* states only. The orbitals ϕ_i and their energies ε_i follow from the one–particle Schrödinger equation

$$\hat{h}^0 \phi_i = \left(-\frac{1}{2}\nabla^2 + v(\mathbf{r})\right)\phi_i = \varepsilon_i \phi_i. \qquad (2.24)$$

Furthermore, for a non–interacting system the Hohenberg–Kohn functional F_{HK}^0 is known, since it only consists of the kinetic energy

$$T^0[\rho] = \sum_i^{occ} \langle \phi_i | -\frac{1}{2}\nabla^2 | \phi_i \rangle. \qquad (2.25)$$

Now the actual trick of Kohn and Sham was to implement the functional $T^0[\rho]$, defined for the interaction-free auxiliary system, into the Hohenberg–Kohn functional F_{HK} of an *interacting* system the following way:

$$\begin{aligned} F_{HK}[\rho] &= T[\rho] + W[\rho] \\ &= T^0[\rho] + J[\rho] + E_{xc}[\rho] \\ \text{with:}\quad J[\rho] &= \frac{1}{2}\iint \frac{\rho(\mathbf{r})\rho(\mathbf{r}')}{|\mathbf{r}-\mathbf{r}'|}d\mathbf{r}d\mathbf{r}' \\ E_{xc}[\rho] &= T[\rho] + W[\rho] - (T^0[\rho] + J[\rho]) \end{aligned} \qquad (2.26)$$

2.3. APPROXIMATIONS

$J[\rho]$ is called the Hartree energy or Coulomb energy and it is known from Hartree–Fock and Thomas–Fermi theory [79, 78]. It represents the classical electrostatic repulsion of the electrons. The non–classical interactions are described by the so called *exchange–correlation functional* $E_{\mathrm{xc}}[\rho]$, which is unknown today and has to be approximated in practice.

Using functional $T^0[\rho]$ in 2.26 introduces the one–particle orbitals ϕ_i to the description of the interacting system. The orbitals are determined by applying the Ritz variational principle to the energy functional $E[\rho]$ that is defined in Eqs. 2.20 and 2.26 and the density is given by Eq. 2.23. To ensure the orthonormality of the orbitals $\langle \phi_k | \phi_l \rangle = \delta_{kl}$ the method of Lagrange multipliers is used:[3]

$$\frac{\delta}{\delta \phi_i^*} \left(E[\rho] - \sum_{kl} \varepsilon_{kl} \left(\langle \phi_k | \phi_l \rangle - \delta_{kl} \right) \right) = 0. \tag{2.27}$$

After a unitary transformation that diagonalizes the matrix of Lagrange multipliers $\varepsilon_{kl} \to \varepsilon_i$, the orbitals ϕ_i and their energies ε_i are

$$\hat{h}\phi_i = \left(-\frac{1}{2}\nabla^2 + v_{\mathrm{eff}}(\mathbf{r}) \right) \phi_i = \varepsilon_i \phi_i$$

$$\text{with:} \quad v_{\mathrm{eff}}(\mathbf{r}) = v(\mathbf{r}) + v_{\mathrm{H}}(\mathbf{r}) + v_{\mathrm{xc}}(\mathbf{r})$$

$$v_{\mathrm{H}}(\mathbf{r}) = \int \frac{\rho(\mathbf{r}')}{|\mathbf{r}-\mathbf{r}'|} d\mathbf{r}'$$

$$v_{\mathrm{xc}}(\mathbf{r}) = \frac{\delta E_{\mathrm{xc}}[\rho]}{\delta \rho(\mathbf{r})} \tag{2.28}$$

This is the Kohn–Sham equation, the heart of modern DFT calculations. It is a non–linear equation since the effective potential v_{eff} depends on the charge density and thus on the orbitals. Therefore it has to be solved self–consistently by iteration. The total electronic energy is not just the sum of the one–particle energies but

$$E[\rho] = \sum_i^{\mathrm{occ}} \varepsilon_i - J[\rho] + E_{\mathrm{xc}}[\rho] - \int v_{\mathrm{xc}}(\mathbf{r})\rho(\mathbf{r}) d\mathbf{r}. \tag{2.29}$$

The Kohn-Sham equation 2.28 is equivalent to equation 2.24 for a system of non–interacting electrons in an external potential $v(\mathbf{r}) = v_{\mathrm{eff}}(\mathbf{r})$. This formal equivalence is the reason why the wave function of the interacting system can be expressed as a Slater determinant and the density by Eq. 2.23. In other words, the interacting system is "mapped" onto an interaction–free auxiliary system. This is done in such a way that in the ground state the density of the interacting system in the external

---
[3]Furthermore we use $\frac{\delta E[\rho]}{\delta \phi_j^*(x)} = \int dx' \frac{\delta E[\rho]}{\delta \rho(x')} \frac{\delta \rho(x')}{\delta \phi_j^*(x)}$.

potential $v(\mathbf{r})$ is the same as the ground state density of a non–interacting auxiliary system in the external potential $v_{\text{eff}}(\mathbf{r})$. So the Kohn-Sham method does actually not describe the real system of interacting electrons but a system of independent particles in the auxiliary system. The latter experience the many–body effects only via the effective potential in an averaged manner. However, the self-consistency cycle adjusts these particles such that in the ground state their properties are on the average the same as the ones of the interacting electrons. Therefore, the Kohn–Sham eigenenergies ε_i and eigenstates ϕ_i (belonging to the particles of the auxiliary system) do not have a well defined physical meaning. However, they are often used to estimate the physical one–particle energies and experience has shown that on a qualitative level a comparison with experiment is indeed possible.

The remaining problem of DFT is to specify the exchange–correlation functional $E_{\text{xc}}[\rho]$. This is by no means trivial. However, several functionals have been developed during the last decades. The most common ones belong to the classes of the *local density approximation* (LDA) [60, 61, 62] and the *generalized gradient approximation* (GGA) [63, 64, 65].

The LDA is related to the Thomas–Fermi theory [76, 77], an early variant of DFT that provides explicit expressions of density functionals from considering the homogeneous electron gas. In the LDA the exchange–correlation functional is given by

$$E_{\text{xc}}^{\text{LDA}}[\rho] = \int e_{\text{xc}}(\rho(\mathbf{r})) \, \rho(\mathbf{r}) d\mathbf{r}, \qquad (2.30)$$

where e_{xc} is the exchange–correlation energy per electron of the uniform electron gas. It can be decomposed into the exchange and correlation energy $e_{\text{xc}} = e_{\text{x}} + e_{\text{c}}$. The exchange energy is given by $e_{\text{x}}(\rho) = -C_{\text{x}} \, \rho^{1/3}$, with $C_{\text{x}} = 3/4(3/\pi)^{1/3}$, and there are no general expression for the correlation energy. For e_{c} one can use, for example, parameterizations of the correlation energies of the uniform electron gas determined in quantum Monte Carlo simulations [81, 62]. The approximation is called *local* density approximation because the functional depends only upon the density at the coordinate where the functional is evaluated. Although derived from the homogeneous electron gas, the LDA gives surprisingly good results for ground state properties of atoms, molecules, and solids, that have rather inhomogeneous electron densities. A well known drawback is its tendency to overbind, i.e., the LDA bond lengths are too small and the binding energies are too big.

A natural way to improve upon the LDA is to take into account the inhomogeneity of the electron density. In the GGA the exchange–correlation functional does not only depend on the density but also on the gradient of the density at each point $\mathbf{r}$. The most general form of the GGA exchange–correlation functional is

$$E_{\text{xc}}^{\text{GGA}}[\rho] = \int f(\rho(\mathbf{r}), \nabla \rho(\mathbf{r})) d\mathbf{r}. \qquad (2.31)$$

2.3. APPROXIMATIONS

A multitude of different realizations of the function $f(\rho(\mathbf{r}), \nabla\rho(\mathbf{r}))$ can be found in the literature [63, 64, 65]. The GGA significantly improves upon the LDA in many respects, for example it significantly reduced the above mentioned overbinding.

Experience has shown that both approximations describe ground state properties (e.g. the total energies E, the charge density $\rho(\mathbf{r})$, etc.) well but do not correctly describe, for example, one–particle energies, band gaps in semiconductors and insulators, excited states, localized d or f electrons, strong electronic correlations, or long–ranged interactions as van der Waals forces or interactions via Hydrogen bonds [66, 67]. However, systems with rather delocalized sp valence electrons, that we are considering here, are well described by the LDA or the GGA.

2.3.3 Density Functional Perturbation Theory

Perturbation theory (PT) allows to find an approximate solution to a problem which cannot be solved exactly, by starting from the exact solution of a related problem. Thus PT allows to calculate physical quantities that are not accessible otherwise.

Sometimes PT is also used to solve a problem more *efficiently* than alternative methods could do. This for example is the case in the calculation of vibrational properties. In Sec. 2.3.1 we described how the Hellmann–Feynman theorem is used to determine the Hesse matrix (see Eq. 2.12). This approach requires the knowledge of the electronic ground state charge density $\rho(\mathbf{r})$ as well as its first derivative $\partial\rho(\mathbf{r})/\partial R_{\beta\nu}$ with respect to an atomic displacement. The former can readily be obtained from DFT calculations and the latter from density functional perturbation theory (DFPT), as will be described now.

Let us recall the principles of perturbation theory: We consider an unperturbed system that is described by the Hamiltonian $\hat{h}^{(0)}$ and whose ground state $\phi^{(0)}$ and ground state energy $\varepsilon^{(0)}$ are known. Now the system is subject to a *small* perturbation by the potential $\lambda\Delta v$ such that $\hat{h} = \hat{h}^{(0)} + \lambda\Delta v$, where λ is a small parameter. Since the perturbation is small we assume that the ground state of $\hat{h}$ is similar to $\phi^{(0)}/\varepsilon^{(0)}$ with some slight corrections. Therefore, the eigenstates ϕ and energies ε of $\hat{h}$ are expressed as a power series in λ: $\phi = \sum_j \lambda^j \phi^{(j)}$ and $\varepsilon = \sum_j \lambda^j \varepsilon^{(j)}$. And the leading (zeroth order) terms of that power series are $\phi^{(0)}$ and $\varepsilon^{(0)}$. This ansatz leads to a set of equations that determine the corrections $\phi^{(j)}/\varepsilon^{(j)}$ for every order j in λ. In first order, the changes in the wave function, energy, potential, etc. are linearly proportional to the magnitude of Δv. That is why first–order perturbation theory is also called *linear response* (LR).

In DFPT the *unperturbed* system, i.e., the zeroth order, is described by a DFT Hamiltonian

$$\hat{h}^{(0)} = -\frac{1}{2}\nabla^2 + v_{\text{eff}}^{(0)}. \qquad (2.32)$$

Its ground state in terms of the orbitals $\phi_i^{(0)}$ and the energies $\varepsilon_i^{(0)}$ is assumed to be known. The effective potential $v_{\text{eff}}^{(0)} = v^{(0)} + v_{\text{H}}(\rho^{(0)}) + v_{\text{xc}}(\rho^{(0)})$ is the external potential $v^{(0)}$, screened by the Hartree and exchange–correlation interactions. The latter two are explicit functions of the ground state charge density $\rho^{(0)} = \sum_i |\phi_i^{(0)}|^2$. The perturbing potential Δv is a correction to $v^{(0)}$

$$v = v^{(0)} + \lambda \Delta v. \tag{2.33}$$

So in the spirit of DFT, both are *external potentials* and are thus subject to screening. With this in mind we can set up the following first–order perturbation series:

$$\begin{aligned} v_{\text{eff}} &= v_{\text{eff}}^{(0)} + \lambda v_{\text{eff}}^{(1)} \\ \phi_i &= \phi_i^{(0)} + \lambda \phi_i^{(1)} \\ \varepsilon_i &= \varepsilon_i^{(0)} + \lambda \varepsilon_i^{(1)} \\ \rho &= \rho^{(0)} + \lambda \rho^{(1)} \end{aligned} \tag{2.34}$$

Inserting this into Eqs. 2.28 and 2.23 leads for the first order in λ:

$$\left(\hat{h}^{(0)} - \varepsilon_i^{(0)}\right)\phi_i^{(1)} = -\left(v_{\text{eff}}^{(1)} - \varepsilon_i^{(1)}\right)\phi_i^{(0)} \tag{2.35}$$

$$\text{with:} \quad v_{\text{eff}}^{(1)}(\mathbf{r}) = \Delta v(\mathbf{r}) + \int \frac{\rho^{(1)}(\mathbf{r}')}{|\mathbf{r}-\mathbf{r}'|}d\mathbf{r}' + \int \frac{\delta v_{\text{xc}}(\mathbf{r})}{\delta \rho(\mathbf{r}')}\rho^{(1)}(\mathbf{r}')d\mathbf{r}' \tag{2.36}$$

$$\text{and:} \quad \rho^{(1)}(\mathbf{r}) = 2\text{Re}\sum_i^{\text{occ}} \left(\phi_i^{(0)}(x)\right)^* \phi_i^{(1)}(x). \tag{2.37}$$

These are the linear response equations. The index i again represents a complete set of one–particle quantum numbers (including spin) and the summation in Eq. 2.37 is over the occupied states only. Equation 2.35 is a well–known expression from perturbation theory. In the context of atomic or solid state physics it is called *Sternheimer* equation [82]. It determines the first–order change of the Kohn–Sham orbitals $\phi_i^{(1)}$ and energies $\varepsilon_i^{(1)} = \langle \phi_i^{(0)} | v_{\text{eff}}^{(1)} | \phi_i^{(0)} \rangle$ due to the screened perturbing potential $v_{\text{eff}}^{(1)}$. Because the linear response of the system depends only on the component of the perturbation that couple the occupied states with the unoccupied ones [47], Eq. 2.35 is rewritten

$$\left(\hat{h}^{(0)} - \varepsilon_i^{(0)}\right)\phi_i^{(1)} = -\hat{P}_{\text{u}}\, v_{\text{eff}}^{(1)} \phi_i^{(0)}$$

$$\text{with:} \quad \hat{P}_{\text{u}} = 1 - \sum_i^{\text{occ}} |\phi_i^{(0)}\rangle\langle\phi_i^{(0)}|. \tag{2.38}$$

Here $\hat{P}_{\text{u}}$ is the projection operator onto the unoccupied–state manifold. Similar to the unperturbed effective potential $v_{\text{eff}}^{(0)}$, its first–order correction $v_{\text{eff}}^{(1)}$ (Eq. 2.36) contains

2.4. PERIODIC SOLIDS

a Hartree and an exchange–correlation term that screen the perturbing potential Δv. The latter two contributions depend on the first–order change in the charge density $\rho^{(1)}$. It is defined in Eq. 2.37 as a sum over the occupied states. Here we encounter the advantage of the DFPT: The linear response of the system due to an external perturbation can be determined from the knowledge of the occupied states only. The standard equations of perturbation theory, in turn, involve summations over unoccupied states which is much more computationally demanding that the present method.

As the potential $v_{\text{eff}}^{(1)}$ in Eq. 2.36 depends linearly on $\rho^{(1)}$, and $\rho^{(1)}$ depends linearly on $\phi_i^{(1)}$, Eqs. 2.35 to 2.37 can be cast into a generalized linear problem (in contrast to the Kohn–Sham equations, which are non–linear). Whether this large linear system is better solved directly (similar to a set of coupled linear equations) or by the self–consistent solutions of the smaller linear system (Eqs. 2.35 to 2.37) is a matter of computational strategy [47].

Our original motivation to use DFPT is the calculation of vibrational properties. In this case the perturbing potential is $\Delta v = \partial v^{(0)}/\partial R_{\alpha\mu}$, which is a simple analytical derivative of the external potential $v^{(0)}$ (defined in Eq. 2.17). Then the first–order change in the charge density is $\partial \rho^{(0)}/\partial R_{\alpha\mu} = \rho^{(1)}$. This quantity and the unperturbed charge density $\rho^{(0)}$ allow to calculate the Hesse matrix $H_{\alpha\mu,\beta\nu}^{\text{HF}}$ via Eq. 2.12 from the Hellman-Feynman theorem, and thus to obtain the vibrational properties of molecules or solids in the harmonic approximation.

2.4 Periodic Solids

In this thesis density functional theory and density functional perturbation theory are applied to *periodic* solids, which are characterized by long–range periodic order. This is in contrast to *quasi crystalline* solids, which are long–range ordered but in a complicated non–periodic way, or *amorphous* solids, where long–range order is absent.

The atomic structure of a periodic solid can be described by a *Bravais lattice*, where every *unit cell* contains one or more *basis atoms*. A Bravais lattice are all points $\mathbf{R}$ in space with

$$\mathbf{R} = n_1 \mathbf{a}_1 + n_2 \mathbf{a}_2 + n_3 \mathbf{a}_3, \quad n_i \in \mathbb{Z}. \tag{2.39}$$

The three vectors $\mathbf{a}_i$ are called *primitive vectors*. The parallelepiped that is spanned by them is called *primitive unit cell* and the vector $\boldsymbol{\tau}_\alpha$ defining the positions of atom α within the unit cell is called *basis vector*. Every real space Bravais lattice can be associated with a dual Bravais lattice of wave vectors called *reciprocal lattice*

$$\mathbf{K} = m_1 \mathbf{b}_1 + m_2 \mathbf{b}_2 + m_3 \mathbf{b}_3, \quad m_j \in \mathbb{Z}$$
$$\text{where:} \quad \mathbf{a}_i \cdot \mathbf{b}_j = 2\pi \delta_{ij}. \tag{2.40}$$

The unit cell of the reciprocal lattice that lies symmetrically around one lattice point **K** and contains all points that are closer to **K** than to any other lattice point is called *first Brillouin zone*. For detailed information see [75].

2.4.1 Energy Band Model

The electronic properties of periodic solids are described with the energy band model. Here we examine the behavior of a system of electrons in a periodic potential

$$\hat{h}\phi_{\mathbf{k}n} = \left(-\frac{1}{2}\nabla^2 + v_{\text{eff}}(\mathbf{r})\right)\phi_{\mathbf{k}n} = \varepsilon_{\mathbf{k}n}\phi_{\mathbf{k}n}$$
$$\text{with:} \quad v_{\text{eff}}(\mathbf{r}) = v_{\text{eff}}(\mathbf{r}+\mathbf{R}). \tag{2.41}$$

The potential v_{eff} is called the *crystal potential* and the lattice vector **R** is defined in Eq. 2.39. The crystal potential is generated by the periodic lattice of the nuclei and also includes the interactions between the electrons as a mean field. Because of the periodicity of v_{eff}, the wave function is

$$\phi_{\mathbf{k}n}(\mathbf{r}) = e^{i\mathbf{k}\cdot\mathbf{r}} u_{\mathbf{k}n}(\mathbf{r})$$
$$\text{with:} \quad u_{\mathbf{k}n}(\mathbf{r}) = u_{\mathbf{k}n}(\mathbf{r}+\mathbf{R}), \tag{2.42}$$

i.e., $\phi_{\mathbf{k}n}$ is a product of a plane wave and a function $u_{\mathbf{k}n}$ that has the periodicity of the lattice. This is *Bloch's theorem* (for proofs see [75]). Equations 2.41 and 2.42 define the energy band model. The energies $\varepsilon_{\mathbf{k}n}$ are called electronic *band structure* and the states $\phi_{\mathbf{k}n}$ are called *Bloch states*, where the *band index* n and the wave vector **k** are quantum numbers.

Using the Bloch state 2.42 in Eq. 2.41 yields a relation that determines the functions $u_{\mathbf{k}n}$, which obey the periodic boundary condition $u_{\mathbf{k}n}(\mathbf{r}) = u_{\mathbf{k}n}(\mathbf{r}+\mathbf{R})$. So the Bloch theorem allows to reduce the treatment of an infinite lattice to calculations of $u_{\mathbf{k}n}$ within a single primitive unit cell. Here the wave vector **k** will merely be a parameter, and for very **k**, the periodic boundary condition leads to a set of *discrete* energies that are labeled with the band index n. Furthermore, it holds $\phi_{\mathbf{k}n} = \phi_{(\mathbf{k}+\mathbf{K})n}$ and $\varepsilon_{\mathbf{k}n} = \varepsilon_{(\mathbf{k}+\mathbf{K})n}$ [75], where **K** is a vector of the reciprocal lattice (Eq. 2.40). In other words, the Bloch states and the band structure are periodic in **k**. Therefore the wave vector **k** can be restricted to within one primitive unit cell of the reciprocal lattice; this is usually the first Brillouin zone.

Overall, Eq. 2.41 describes a system of independent particles that experience interactions only via the effective potential v_{eff}. It is similar to the non–interacting auxiliary system in DFT. Therefore the Kohn–Sham equation 2.28 together with the Bloch relation 2.42 allow to calculate the band structure of a solid.

2.4. PERIODIC SOLIDS

Band Velocity

The band velocity $\mathbf{v}_{\mathbf{k}n}$ is defined as the gradient of the band structure $\varepsilon_{\mathbf{k}n}$ with respect to the wave vector $\mathbf{k}$

$$\mathbf{v}_{\mathbf{k}n} = \nabla_{\mathbf{k}}\, \varepsilon_{\mathbf{k}n}. \tag{2.43}$$

It describes the mean velocity of an electron in the state $\{\mathbf{k}n\}$.

Electronic Density of States

The total electronic density of states of a solid is given by

$$D(\varepsilon) = \sum_{\mathbf{k}n} \delta(\varepsilon - \varepsilon_{\mathbf{k}n}). \tag{2.44}$$

$D(\varepsilon)d\varepsilon$ quantifies the number of electronic states that exist in an energy range of width $d\varepsilon$ at the energy ε.

Fermi Energy and Fermi Surface

Since electrons are Fermions, every one–particle state can be occupied by maximally two electrons of opposite spin. In a solid at absolute zero temperature, all states starting from the one(s) with the lowest energy up to the Fermi level are occupied. Therefore the *Fermi level* or *Fermi energy* ε_F is the energy of the highest occupied electronic state. It is defined by

$$N = 2 \int_0^{\varepsilon_\mathrm{F}} D(\varepsilon)\, d\varepsilon, \tag{2.45}$$

where N is the number of electrons per primitive unit cell and the factor 2 stems from spin degeneracy. The *Fermi surface* is a constant–energy surface in $\mathbf{k}$–space defined by

$$\varepsilon_{\mathbf{k}n} = \varepsilon_\mathrm{F}. \tag{2.46}$$

Brillouin Zone Averages

We are usually considering an infinite solid. In that case the band structure and the Bloch states are continuous functions in $\mathbf{k}$ [75]. And the summation over $\mathbf{k}$–vectors $\sum_\mathbf{k}$ in Eq. 2.44 is a short notation for

$$\sum_\mathbf{k} \stackrel{\mathrm{def}}{=} \frac{1}{\Omega} \int_{\mathrm{BZ}} d\mathbf{k}. \tag{2.47}$$

Definition 2.47 represents an integral over the first Brillouin zone (or any other primitive unit cell in reciprocal space), normalized by its volume Ω. This is a Brillouin zone (BZ) average. In this thesis symbolic summation over **k**–vectors will always represent a BZ average.

2.4.2 Lattice Vibrations

In Sec. 2.3.1 we described how atomic vibrations are treated in the adiabatic approximation. In a periodic solid the equilibrium position $\mathbf{R}_\alpha^I$ of an atom is $\mathbf{R} + \boldsymbol{\tau}_\alpha$, where **R** is a Bravais lattice vector according to Eq. 2.39 and $\boldsymbol{\tau}_\alpha$ defines the equilibrium positions of atom α within the unit cell; each unit cell contains Γ atoms ($\alpha = 1 \ldots \Gamma$). The position of a displaced atom is then

$$\mathbf{r}_\alpha(\mathbf{R}) = \mathbf{R} + \boldsymbol{\tau}_\alpha + \mathbf{u}_\alpha(\mathbf{R}), \qquad (2.48)$$

and the Cartesian displacement $\mathbf{u}_\alpha(\mathbf{R})$ of atom α can differ from cell to cell. Now the Hesse matrix is

$$H_{\alpha\mu,\beta\nu}(\mathbf{R}_m, \mathbf{R}_n) = \frac{\partial^2 E}{\partial r_{\alpha\mu}(\mathbf{R}_m)\,\partial r_{\beta\nu}(\mathbf{R}_n)}, \qquad (2.49)$$

where $\mu, \nu = x, y, z$ are the Cartesian components. Due to the periodicity of the Bravais lattice, all unit cells are equivalent. Therefore it holds

$$H(\mathbf{R}_m, \mathbf{R}_n) = H(\mathbf{R}_m - \mathbf{R}_n) = H(\mathbf{R}). \qquad (2.50)$$

Analogous to Sec. 2.3.1 mass–weighted displacements $\mathbf{w}_\alpha(\mathbf{R}) = \sqrt{M_\alpha}\mathbf{u}_\alpha(\mathbf{R})$ are now introduced and the kinetic energy and harmonic potential operators become

$$\hat{T}_N = -\frac{1}{2}\sum_{\mathbf{R}}\sum_{\alpha=1}^{\Gamma}\sum_{\mu=x,y,z}\left(\frac{\partial}{\partial w_{\alpha\mu}(\mathbf{R})}\right)^2 \qquad (2.51)$$

$$\hat{W}_{\text{harm}} = \frac{1}{2}\sum_{\mathbf{R}_m,\mathbf{R}_n}\sum_{\alpha,\beta=1}^{\Gamma}\sum_{\mu,\nu=x,y,z} H'_{\alpha\mu,\beta\nu}(\mathbf{R}_m - \mathbf{R}_n)\, w_{\alpha\mu}(\mathbf{R}_m)\, w_{\beta\nu}(\mathbf{R}_n)$$

$$\text{with: } H'_{\alpha\mu,\beta\nu}(\mathbf{R}) = \frac{1}{\sqrt{M_\alpha M_\beta}} H_{\alpha\mu,\beta\nu}(\mathbf{R}). \qquad (2.52)$$

Inserting these expressions into Hamiltonian 2.5 leads to a set of equations that couple the movements of the Γ atoms of one unit cell among each other and to the atoms of all other unit cells. As we are considering an infinite solid the number of coupled equations is infinite. However, the periodicity of the lattice (Eq. 2.50) can

2.4. PERIODIC SOLIDS

be used to decouple this system. In the following ansatz for the (mass–weighted) atomic displacements

$$w_{\alpha\mu}(\mathbf{R}) = \epsilon_{\alpha\mu}(\mathbf{q}) \, e^{i\mathbf{q}\cdot\mathbf{R}-i\omega(\mathbf{q})t} \tag{2.53}$$

the dependence of the displacements on the unit cell is represented in the spatial part of a wave solution and the polarizations $\epsilon_{\alpha\mu}$ and the vibrational frequency ω depend on the wave vector $\mathbf{q}$. This ansatz leads to an eigenvalue equation (similar to Eq. 2.16)

$$D(\mathbf{q})\epsilon_j(\mathbf{q}) = \omega_j^2(\mathbf{q})\epsilon_j(\mathbf{q}) \tag{2.54}$$

for every wave vector $\mathbf{q}$. $D(\mathbf{q})$ is the Fourier transform of the Hesse matrix

$$D_{\alpha\mu,\beta\nu}(\mathbf{q}) = \sum_{\mathbf{R}} e^{-i\mathbf{q}\cdot\mathbf{R}} H'_{\alpha\mu,\beta\nu}(\mathbf{R}) \tag{2.55}$$

and it is called the *dynamical matrix*. D is a $3\Gamma \times 3\Gamma$ matrix and the polarization vectors ϵ_j are 3Γ-dimensional vectors. Solving Eq. 2.54 yields $j = 1\ldots 3\Gamma$ eigenvalues ω_j^2 and eigenvectors ϵ_j. We find that the wave ansatz 2.53 reduces the infinite number of equations to 3Γ equations, that have to be solved for every wave vector $\mathbf{q}$. The dynamical matrix is periodic in reciprocal space and therefore $\mathbf{q}$ can be restricted to within one primitive unit cell of the reciprocal lattice (usually the first Brillouin zone). Thus $\mathbf{q}$ has the same properties as the wave vector $\mathbf{k}$ in the energy band model; they are defined on the same domain.

The function $\omega_j(\mathbf{q})$ is called *phonon dispersion* and represents the vibrational frequency of a eigenmode j with a wave vector $\mathbf{q}$. To obtain $\omega_j(\mathbf{q})$ we treated Hamiltonian 2.5 classically. However, a quantum mechanical treatment shows that the wave solutions according to equation 2.53 can be quantized. The bosonic excitation quanta of these lattice vibrations are called *phonons* [75, 47, 68].

In the following the phonon dispersion and the polarization vectors will be referred to as

$$\begin{aligned} \omega_{\mathbf{q}\nu} &= \omega_j(\mathbf{q}) \\ \epsilon_{\mathbf{q}\nu,\alpha\mu} &= [\epsilon_j(\mathbf{q})]_{\alpha\mu} \end{aligned} \tag{2.56}$$

where the phonon mode number is $j = \nu$. The matrix of polarizations vectors $\epsilon_{\mathbf{q}\nu,\alpha\mu}$ represents a transformation from the system of Cartesian coordinates $(\alpha\mu)$ to *normal coordinates* $(\mathbf{q}\nu)$, which are dynamically independent of one another.

Phononic Density of States

The phononic density of states is defined analogous to the electronic density of states (see Eq. 2.44)

$$F(\omega) = \sum_{\mathbf{q}\nu} \delta(\omega - \omega_{\mathbf{q}\nu}), \tag{2.57}$$

where the sum over the wave vector **q** again represents a Brillouin zone average (see Sec. 2.4.1).

2.4.3 Electron–Phonon Coupling

Electrons at the Fermi level are sensitive to the time dependence of the lattice vibrations. This leads to interactions between electrons and phonons which give rise to physical phenomena such as electrical and thermal resistivities and superconductivity.

In Sec. 2.3.3 we treated atomic vibrations as first–order perturbation to the electronic ground state given by DFT. The perturbing potential was time–independent. However the interactions of electrons and phonons are time–dependent phenomena and can be calculated with time–dependent perturbation theory. Its first–order approximation for a periodic perturbation is the Fermi "golden rule". Here the central quantity is the electron–phonon matrix element

$$g^{\nu}_{(\mathbf{k+q})n',\mathbf{k}n} = \langle(\mathbf{k+q})n'|v^{(1)}_{\text{eff},\mathbf{q}\nu}|\mathbf{k}n\rangle \tag{2.58}$$

$$\text{with:} \quad v^{(1)}_{\text{eff},\mathbf{q}\nu} = \sum_{\alpha=1}^{\Gamma}\sum_{\mu=x,y,z} \frac{1}{\sqrt{2M_\alpha \omega_{\mathbf{q}\nu}}} \epsilon_{\mathbf{q}\nu,\alpha\mu} \, v^{(1)}_{\text{eff},\alpha\mu}, \tag{2.59}$$

where $v^{(1)}_{\text{eff},\alpha\mu}$ is the self–consistent first–order change of the Kohn–Sham potential (given by Eq. 2.36) due to a perturbation $\Delta v = \partial v/\partial R_{\alpha\mu}$, and $\epsilon_{\mathbf{q}\nu,\alpha\mu}$ is the matrix of polarizations vectors (see Eq. 2.56), and $1/\sqrt{2M_\alpha \omega_{\mathbf{q}\nu}}$ is called the zero–point phonon amplitude. Equation 2.59 is a transformation from the Cartesian coordinates of the displacements to the normal coordinates of the phonon mode ($\mathbf{q}\nu$) amended by the zero–point amplitude. The electron–phonon matrix element Eq. 2.58 represents the transition amplitude for an electron or hole in the band state $|\mathbf{k}n\rangle$ to be scattered to the state $|(\mathbf{k+q})n'\rangle$ by absorption or emission of a phonon ($\mathbf{q}\nu$). Its absolute square is the corresponding transition probability.

This scattering process crucially depends on the density of initial and final states. If the phonon energies are small compared to the electronic energies it is a good approximation to assume that only the states at the Fermi level ε_F contribute to the electron–phonon interactions. The available phase space for electron or hole scattering across the Fermi surface is described by the so called nesting function

$$\begin{aligned}\chi_\mathbf{q} &= \sum_{\mathbf{k}nn'} \delta(\varepsilon_{\mathbf{k}n}-\varepsilon_\text{F})\delta(\varepsilon_{(\mathbf{k+q})n'}-\varepsilon_\text{F}) \\ &= \frac{1}{\Omega}\sum_{nn'}\oint_L \frac{dL_\mathbf{k}}{|\mathbf{v}_{\mathbf{k}n}\times\mathbf{v}_{(\mathbf{k+q})n'}|}.\end{aligned} \tag{2.60}$$

$\chi_\mathbf{q}$ can be expressed as closed line integral over the intersections $L_\mathbf{k}$ of an undisplaced FS and one that is displaced by the vector $\mathbf{q}$; $\mathbf{v}_{\mathbf{k}n}$ is the band velocity defined in

2.4. PERIODIC SOLIDS

Eq. 2.43. It holds $\sum_\mathbf{q} \chi_\mathbf{q} = [D(\varepsilon_F)]^2$. From Eq. 2.60 it is further obvious that $\chi_{\mathbf{q}=0}$ is divergent.

If we now combine the electron–phonon matrix element and the nesting function, we obtain a quantity $\beta_{\mathbf{q}\nu}$ that describes how electrons couple to the phonon mode ($\mathbf{q}\nu$)

$$\beta_{\mathbf{q}\nu} = \sum_{\mathbf{k}nn'} \left|g^{\nu}_{(\mathbf{k+q})n',\mathbf{k}n}\right|^2 \delta(\varepsilon_{\mathbf{k}n} - \varepsilon_F)\delta(\varepsilon_{(\mathbf{k+q})n'} - \varepsilon_F), \quad (2.61)$$

and that is related to a number of physical quantities that characterize electron–phonon coupling. The half–width at half–maximum phonon linewidth $\gamma_{\mathbf{q}\nu}$ and the mode coupling constant $\lambda_{\mathbf{q}\nu}$ are then defined as

$$\gamma_{\mathbf{q}\nu} = 2\pi\, \omega_{\mathbf{q}\nu} \beta_{\mathbf{q}\nu} \quad (2.62)$$

$$\lambda_{\mathbf{q}\nu} = \frac{2}{D(\varepsilon_F)} \frac{\beta_{\mathbf{q}\nu}}{\omega_{\mathbf{q}\nu}}. \quad (2.63)$$

Definition 2.62 is the Fermi "golden rule" [83], and Eq. 2.63 is a phonon mode decomposition of the electron–phonon coupling constant λ, defined below. As $\beta_{\mathbf{q}\nu}$ is inversely proportional to $\omega_{\mathbf{q}\nu}$ (see Eq. 2.59, 2.58, and 2.61) it follows that $\lambda_{\mathbf{q}\nu}$ is inversely proportional to the square of the phonon frequency. So coupling to phonons with a low–frequency results in a large $\lambda_{\mathbf{q}\nu}$. The so called *Eliashberg spectral function* is defined as

$$\begin{aligned}\alpha^2 F(\omega) &= \frac{1}{D(\varepsilon_F)} \sum_{\mathbf{q}\nu} \beta_{\mathbf{q}\nu}\, \delta(\omega - \omega_{\mathbf{q}\nu}) \\ &= \frac{1}{2\pi D(\varepsilon_F)} \sum_{\mathbf{q}\nu} \frac{\gamma_{\mathbf{q}\nu}}{\omega_{\mathbf{q}\nu}} \delta(\omega - \omega_{\mathbf{q}\nu}). \end{aligned} \quad (2.64)$$

This function is the central quantity in *Midgal–Eliashberg theory* [84, 85, 86] that describes the superconducting state of a solid to high accuracy by field–theoretic methods. The **q**–resolved total coupling is obtained by summing over all phonon modes

$$\lambda_\mathbf{q} = \sum_\nu \lambda_{\mathbf{q}\nu}, \quad (2.65)$$

and the total electron–phonon coupling constant $\lambda = \lambda_{\text{tot}}$ (also called *mass–enhancement parameter*) is the BZ average[4] of the $\lambda_\mathbf{q}$

$$\begin{aligned}\lambda &= \sum_\mathbf{q} \lambda_\mathbf{q} = \sum_{\mathbf{q}\nu} \lambda_{\mathbf{q}\nu} \\ &= 2\int_0^\infty \frac{\alpha^2 F(\omega)}{\omega} d\omega. \end{aligned} \quad (2.66)$$

---

[4]We want to remind the reader that throughout this thesis symbolic summation over **k** or **q** vectors represents BZ averaging (see Sec. 2.4.1).

The second line is an relation between $\alpha^2 F(\omega)$ and λ.

2.4.4 Superconductivity

In some metals the electron–phonon interactions mediate an effective attraction between electrons at low temperatures. This leads to the formation of bound states of two electrons – the *Cooper pairs*. The superconducting state is a macroscopic condensate of Cooper pairs and is characterized by exactly zero electrical resistance and the exclusion of the interior magnetic field (the *Meissner effect*). The highest temperature at which superconductivity is possible in a certain material (and without external magnetic fields) is called *critical temperature* T_c. The pairing mechanism based on electron–phonon interactions is well understood. If the coupling constant λ is small the superconducting state can be described with *BCS theory*, named after its founders Bardeen, Cooper, and Schrieffer [87, 88]. However, this theory fails if $\lambda > 0.1 - 0.2$ because the electron–phonon interactions are not treated accurately enough. These problems are overcome in *Midgal–Eliashberg theory* [84, 85, 86], the strong–coupling version of the BCS theory. It treats the time–dependent electron–phonon interactions explicitly and allows to determine the superconducting transition temperature T_c from the knowledge of the spectral function $\alpha^2 F(\omega)$ (defined in Eq. 2.64). This involves the solution of two coupled integral equations, which are called *Eliashberg equations* [89]. To estimate T_c, however, solving the Eliashberg equations can often be circumvented as there are approximate equations for T_c that are based on a few parameters only. For couplings $\lambda < 1$ the McMillan equation [90] provides good estimates for T_c.

$$T_c = \frac{\omega_{\log}}{1.2} \exp\left(-\frac{1.04(1+\lambda)}{\lambda - \mu^*(1+0.62\lambda)}\right) \quad (2.67)$$

Here λ is the total electron–phonon coupling constant given by Eq. 2.66, $\omega_{\log}$ is a characteristic frequency defined in Eq. 2.70, and μ^* is called Coulomb pseudopotential. The latter is an dimensionless empirical parameter of the order 0.1, that is almost constant for the different materials. It describes the effective Coulomb repulsion[5] felt by the electrons.

The parameter $\omega_{\log}$ is a logarithmic moment of the weighting function

$$g(\omega) = \frac{2}{\lambda \omega} \alpha^2 F(\omega), \quad (2.68)$$

which is normalized to one: $\int_0^\infty g(\omega) d\omega = 1$. The coupling–weighted phonon mo-

---

[5] In a typical conventional superconductor the Coulomb repulsion between the electrons is significantly reduced by *retardation effects* [89].

2.5. BASIS SETS

ments are then

$$\omega_n = \left(\int_0^\infty g(\omega)\omega^n d\omega\right)^{\frac{1}{n}} \tag{2.69}$$

$$\omega_{\log} = \exp\left(\int_0^\infty g(\omega)\ln\omega d\omega\right). \tag{2.70}$$

2.5 Basis Sets

In usual implementations of DFT (and DFPT) the wave functions $|\phi\rangle$ are expanded in a basis set $\{\chi_j\}$ of finite size

$$|\phi\rangle = \sum_{i=1}^{M} c_i |\chi_i\rangle. \tag{2.71}$$

This transforms the Kohn–Sham differential equation $(\hat{h} - \varepsilon)|\phi\rangle = 0$ (see Eq. 2.28) into the matrix equation

$$(H - \varepsilon O) \cdot \boldsymbol{c} = \boldsymbol{0}, \tag{2.72}$$

where H and O are the Hamiltonian and overlap matrices

$$H_{ij} = \langle \chi_i | \hat{h} | \chi_j \rangle \tag{2.73}$$
$$O_{ij} = \langle \chi_i | \chi_j \rangle, \tag{2.74}$$

ε is the eigenenergy, and $\boldsymbol{c} = (c_1, \ldots, c_M)$ is the vector of the expansion coefficients c_i. Equation 2.72 can be solved numerically with standard methods of linear algebra. Choosing basis functions that are close to the correct solution, allows to use a small basis set and reduces the computational costs significantly.

In this thesis two kinds of basis sets were used, which will be introduced now.

2.5.1 Linear Muffin Tin Orbitals

Linear muffin tin orbitals (LMTOs) were developed by O. K. Andersen [91, 92] for the energy band model (see Sec. 2.4.1) but are not restricted to it. Similar to the *(linear) augmented plain wave* method (LAPW/APW) [93, 92] and the method of *Korringa, Kohn, Rostocker* (KKR) [94, 95], LMTO is based on the *partial wave* approach. Here space is separated into different regions and solutions of Schrödinger equation in the different regions (the partial waves) are matched at their boundaries.

Muffin Tin Potential

The *muffin tin potential* approximation was suggested by Slater [93] in conjuction with the APW method. It is based on the idea that the primary difference between the electronic states in an atom and in a solid are the different boundary conditions and that the corresponding change in the potential is less important. Therefore the effective one–particle potential $v_{\text{eff}}(\mathbf{r})$ (the Kohn–Sham potential in Eq. 2.28 or the crystal potential in Eq. 2.41) is approximated by by a simplified muffin tin (MT) potential

$$v^{\text{MT}}(\mathbf{r}) = \begin{cases} v(r) & r \leq S \\ v_{\text{MTZ}} & r > S \end{cases}. \quad (2.75)$$

For an atom at site $\mathbf{R}$ it is spherical symmetric inside an atom–centered sphere of radius S, called the muffin tin sphere, and has a constant value v_{MTZ}, the muffin tin zero, between the spheres (the *interstitial* region). Around the point $\mathbf{R}$ we use spherical coordinates where $r = |\mathbf{r} - \mathbf{R}|$ and $\hat{r} = (\theta, \phi)$ is the angular part of $\mathbf{r} - \mathbf{R}$.

A *single* MT potential is spherical symmetric everywhere in space and therefore its solutions – let us call them *single muffin tin orbitals* (SMTOs) – are given by

$$\chi_L^{\text{SMTO}}(\varepsilon, \mathbf{r}) = i^l Y_L(\hat{r}) \Psi_l(\varepsilon, r). \quad (2.76)$$

The angular part is described by spherical harmonics $Y_L(\hat{r})$, where the phase factor i^l is chosen for convenience, and $L = lm$ represents the quantum numbers of angular momentum. The radial part $\Psi_l(\varepsilon, r)$ obeys the radial Schrödinger equation[6]

$$\left(-\frac{d^2}{dr^2} + \frac{l(l+1)}{r^2} + v(r) - \varepsilon \right) r \Psi_l(\varepsilon, r) = 0, \quad (2.77)$$

and can be found by numerical integration, where it parametrically depends on the one–particle energy ε. For $r > S$ the potential is constant (v_{MTZ}) and the solutions of Eq. 2.77,[7] called the *tail* or *envelope* of the orbital, are spherical Bessel j_l and Neumann functions n_l. They describe spherical waves of fixed angular momentum (quantum number l) with a kinetic energy

$$\kappa^2 = \varepsilon - v_{\text{MTZ}}, \quad (2.78)$$

and κ^2 is called the tail energy. Now, for positive tail energies the SMTO is

$$\chi_L^{\text{SMTO}}(\varepsilon, \mathbf{r}) = i^l Y_L(\hat{r}) \begin{cases} \Psi_l(\varepsilon, r) & r \leq S \\ \kappa \left[n_l(\kappa r) - \cot(\eta_l) \, j_l(\kappa r) \right] & r > S \end{cases}. \quad (2.79)$$

---

[6] In this section we use $\hbar = 2m = e^2/2 = 1$, which defines atomic Rydberg units.

[7] For a constant potential Eq. 2.77 describes the radial part of the *Helmholtz* equation

$$(\nabla^2 + \kappa^2) \chi(\mathbf{r}) = 0.$$

2.5. BASIS SETS

For $r > S$ we use a linear combination of j_l and n_l, because the two are linearly independent. The prefactor κ (defined in Eq. 2.78) and $\cot(\eta_l)$[8] are chosen such that χ_L^{SMTO} is continuous and differentiable everywhere in space, i.e., the radial solutions inside and outside of the MT sphere are matched smoothly at the sphere boundary $r = S$. If κ^2 is negative, bound states are formed in the MTs and n_l has to be replaced by the spherical Hankel function $n_l - i\, j_l$ in Eqs. 2.79 and 2.80.

For a solid the one–particle potential is approximated by a superposition of MT potentials

$$v_{\text{eff}}(\mathbf{r}) = \sum_{\mathbf{R}} v^{\text{MT}}(\mathbf{r} - \mathbf{R}). \tag{2.81}$$

Now, inside a given MT sphere the radial solution $\Psi_l(\varepsilon, r)$ will overlap with the tails of the orbitals of neighboring atoms. This effect cannot be represented by the SMTOs. Furthermore, χ_L^{SMTO} can have unbound solutions that are not normalizable. Thus SMTOs are not well suited as basis functions for the electronic structure problem.

Muffin Tin Orbitals

However, these difficulties can be remedied by adding spherical Bessel function to the SMTO: $\chi_L^{\text{MTO}} = \chi_L^{\text{SMTO}} + \kappa\, \cot(\eta_l)\, j_l(\kappa r)$. A *muffin tin orbital* (MTO) is then given by [91]

$$\chi_L^{\text{MTO}}(\varepsilon, \mathbf{r}) = i^l Y_L(\hat{r}) \begin{cases} \Psi_l(\varepsilon, r) + \kappa\, \cot(\eta_l)\, j_l(\kappa r) & r \leq S \\ \kappa n_l(\kappa r) & r > S \end{cases}. \tag{2.82}$$

The MTO has the desirable properties that it is continuous and differentiable everywhere in space and normalizable for all values of κ^2. Normalization is possible because the spherical Bessel function j_l is regular at the origin, for bigger r it can compensate the (potentially) divergent part of $\Psi_l(\varepsilon, r)$, and the tail n_l is regular for $r \to \infty$. Furthermore, at a given site $\mathbf{R}$ the tail of a neighboring orbital (spherical Neumann or spherical Hankel function) may be expanded in terms of spherical Bessel functions j_l *inside* the MT sphere at $\mathbf{R}$. Thus the effect of neighboring atoms is taken into account when adding j_l inside the MT sphere. However, in this way the MTO is not a solution of the Schrödinger equation inside the MT sphere anymore. This is only $\Psi_l(\varepsilon, r)$ by virtue of Eq. 2.77. But this property is restored if

---
[8]
$$\cot(\eta_l[\varepsilon, \kappa]) = \frac{n_l(\kappa r)}{j_l(\kappa r)} \cdot \frac{D_l(\varepsilon) - \kappa r\, n_l'(\kappa r)/n_l(\kappa r)}{D_l(\varepsilon) - \kappa r\, j_l'(\kappa r)/j_l(\kappa r)}\bigg|_{r=S}$$

$$\text{with:} \quad D_l(\varepsilon) = \frac{r\Psi_l'(\varepsilon, r)}{\Psi_l(\varepsilon, r)}\bigg|_{r=S}, \quad f'(r) = \frac{\partial f(r)}{\partial r} \tag{2.80}$$

the term $\kappa\cot(\eta_l)j_l(\kappa r)$ is exactly canceled out by the sum of the tails of *all* neighboring atoms. This condition is called *tail cancellation* and is a central part of the MTO formalism [91]. The tail cancellation equation (not given here, see [96]) is an alternative to the approach connected to Eq. 2.72.

Because the MTOs depend on the one–particle energies ε the secular equation of Eq. 2.72

$$|H(\varepsilon) - \varepsilon\, O(\varepsilon)| = 0, \qquad (2.83)$$

has a complicated non–linear energy dependence and is hard to solve in practice.

Linear Muffin Tin Orbitals

In order to obtain an efficient method, it would be desirable to construct orbitals that are energy–independent, i.e., a fixed basis set. To remove the energy dependence of the tails, we now disregard relation 2.78 and consider κ as fixed parameter, independent of ε. In this way the tails become approximate rather than exact solution of the Schrödinger equation in the interstitials. The energy dependence of the radial solutions $\Psi_l(\varepsilon, r)$ inside the MT spheres can be expressed as a first–order Taylor expansion about a fixed energy $\varepsilon_{\nu l}$

$$\Psi_l(\varepsilon, r) = \Psi_{\nu l}(r) \; + \; (\varepsilon - \varepsilon_{\nu l})\, \dot{\Psi}_{\nu l}(r) + \ldots$$
$$\text{with:} \quad \Psi_{\nu l}(r) = \Psi_l(\varepsilon_{\nu l}, r), \qquad \dot{\Psi}_{\nu l}(r) = \left.\frac{\partial \Psi_l(\varepsilon, r)}{\partial \varepsilon}\right|_{\varepsilon=\varepsilon_{\nu l}} \qquad (2.84)$$

In other words $\Psi_l(\varepsilon, r)$ is approximated by a linear combination of the energy–independent function $\Psi_{\nu l}(r)$ and its first energy derivative $\dot{\Psi}_{\nu l}(r)$. In this way the wave functions are correct to first order in $(\varepsilon - \varepsilon_{\nu l})$ and the energies to $(\varepsilon - \varepsilon_{\nu l})^3$, due to the variational principle. $\Psi_{\nu l}(r)$ and $\dot{\Psi}_{\nu l}(r)$ are used to construct a basis for a particular system, that allows to calculate the states in a certain energy window around $\varepsilon_{\nu l}$. In practice this window is sufficiently large to span the occupied part of the valence bands and thus to allow DFT calculations with good accuracy.

To keep the desirable properties of the MTOs (continuity, differentiability, normalizability) we require the LMTOs to have a similar form

$$\chi_L^{\text{LMTO}}(\kappa, \mathbf{r}) = i^l Y_L(\hat{r}) \begin{cases} \Psi_{\nu l}(r) + \kappa\, \cot(\eta_l)\, J_l(\kappa r) & r \leq S \\ \kappa N_l(\kappa r) & r > S \end{cases}, \qquad (2.85)$$

where the spherical Bessel and Neumann functions j_l and n_l are replaced by *augmented* spherical Bessel and Neumann functions J_l and N_l. The latter are constructed in order to make the LMTO (approximately) energy–independent and orthogonal to the core states (such that the eigenvalues of the valence states cannot

2.5. BASIS SETS

erroneously converge to core values). They are

$$J_l(\kappa r) = \begin{cases} -\dot{\Psi}_{\nu l}(r)/\left[\kappa \cot(\eta_l[\varepsilon_{\nu l}])\right] & r \leq S \\ j_l(\kappa r) & r > S \end{cases}, \quad (2.86)$$

$$N_L(\kappa, \mathbf{r} - \mathbf{R}) = \quad (2.87)$$
$$\begin{cases} 4\pi \sum_{L'L''} c_{LL'L''} J_{L'}(\kappa, \mathbf{r} - \mathbf{R}') n_{L''}^*(\kappa, \mathbf{R} - \mathbf{R}') & \begin{cases} |\mathbf{r} - \mathbf{R}'| \leq S \\ \forall \, \mathbf{R}' \neq \mathbf{R} \end{cases} \\ n_L(\kappa, \mathbf{r} - \mathbf{R}) & \text{otherwise} \end{cases}$$

with

$$c_{LL'L''} = \int Y_L(\hat{r}) Y_{L'}^*(\hat{r}) Y_{L''}(\hat{r}) \, d\hat{r} \quad (2.88)$$
$$J_L(\kappa, \mathbf{r}) = i^l Y_L(\hat{r}) J_l(\kappa r) \quad (2.89)$$
$$n_L(\kappa, \mathbf{r}) = i^l Y_L(\hat{r}) n_l(\kappa r). \quad (2.90)$$

Equation 2.88 defines the so called Gaunt coefficients $c_{LL'L''}$. For details about the construction of J_l and N_l see [96].

The LMTO defined above is energy–independent ($\dot{\chi}_L^{\mathrm{LMTO}}(\kappa, \mathbf{r}) = 0$) to first order in $(\varepsilon - \varepsilon_\nu)$. Thus the LMTO secular equation becomes

$$|H - \varepsilon O| = 0. \quad (2.91)$$

The computational cost for solving Eq. 2.91 is orders of magnitude lower than the cost for solving Eq. 2.83. Furthermore, LMTOs allow to express the electronic structure problem in terms of a minimal basis set, i.e., one uses only one basis function for every atomic orbital that is required to describe the free atom. This underlines the efficiency of the LMTO method and thus its practical importance [92, 97, 96].

In the *tight–binding LMTO* (TB-LMTO) scheme the long–ranged LMTOs are transformed (by a unitary transformation) into a tight–binding (TB) representation, were the orbitals are short–ranged (they vanish beyond second–nearest neighbors) and Hamilton and overlap integrals have a two–center from. This allows to formulate a TB theory directly from first principles [98].

Atomic Sphere Approximation

In the *atomic sphere approximation* (ASA) [99] the volume of the MT spheres is chosen to be equal to the atomic volume. For closed–packed crystals (e.g. fcc, bcc, etc.) this leads to a collection of slightly overlapping *atomic spheres* that are space–filling. For open structures *empty spheres* may be introduced in the interstitial region to simulate a closed–packed system. Because of the absence of an interstitial region in the ASA, we can conveniently set the tail energy $\kappa^2 = 0$. This simplifies the LMTOs because for $\kappa \to 0$ the spherical Bessel and Neumann functions are replaced by their simple asymptotes which are proportional to r^l and r^{-l-1}, respectively. The precision of the ASA can be increased by adding so called *combined corrections* terms [100] to the Hamiltonian and overlap matrices.

In this thesis we used the Stuttgart TB-LMTO-ASA program that employs the TB-LMTO basis set and the ASA.

Full Potential LMTO

Above we have shown how the MT potential approximation may be used to construct a fixed basis set of LMTOs tailored individually for each system. However, if accurate total energies, atomic forces (Eq. 2.7), or interatomic force constants (Eq. 2.11) are required, the MT potential approximation is not accurate enough and it becomes necessary to consider the full complexity of the one–particle potential $v_{\text{eff}}(\mathbf{r})$.

In order to do so the unit cell is divided into non–overlapping MT spheres (that are at most touching) and the interstitial region, where no empty spheres are used. The potential and the charge density are expanded in spherical harmonics inside the MT spheres and in plane waves in the interstitials. Inside the MT spheres the LMTO basis functions are constructed from the MT potential, which is the spherical average of the full potential. In order to increase the variational freedom of the wave functions (and thus to increase the numerical accuracy) the LMTO basis function are constructed for multiple κ values. Hence the basis set is no longer minimal.

All this is implemented for example in the program LMTART developed by S. Y. Savrasov [48].

2.5.2 Plain Waves and Pseudopotentials

Expansion in Plain Waves

If we consider a periodic solid, the wave functions are Bloch states according to Eq. 2.42. Since the function $u_{\mathbf{k}n}(\mathbf{r})$ has the periodicity of the lattice, it can be represented as discrete Fourier series

$$u_{\mathbf{k}n}(\mathbf{r}) = \sum_{\mathbf{K}} c_{(\mathbf{k}+\mathbf{K})n} \exp[i\mathbf{K}\cdot\mathbf{r}], \qquad (2.92)$$

where the vectors $\mathbf{K}$ are reciprocal lattice vectors (see Eq. 2.40) and the $c_{(\mathbf{k}+\mathbf{K})n}$ are the expansion coefficients. The Bloch wave function then becomes

$$\phi_{\mathbf{k}n}(\mathbf{r}) = \sum_{\mathbf{K}} c_{(\mathbf{k}+\mathbf{K})n} \exp[i(\mathbf{k}+\mathbf{K})\cdot\mathbf{r}], \qquad (2.93)$$

2.5. BASIS SETS

which is an expansion of $\phi_{\mathbf{k}n}(\mathbf{r})$ in a basis of plane waves. The size of the basis set is defined by the cutoff energy

$$\frac{1}{2}|\mathbf{k}+\mathbf{K}|^2 < E_{\text{cutoff}}. \tag{2.94}$$

All plane waves with a kinetic energy less than E_{cutoff} are included in the basis set. The error induced by the cutoff can be minimized by successively increasing E_{cutoff} until the total energy of the system E is converged to the required accuracy. In this way the basis set can be improved systematically.

In the basis of plane waves the overlap matrix simply is $O_{(\mathbf{k}+\mathbf{K}'),(\mathbf{k}+\mathbf{K})} = \delta_{\mathbf{K}',\mathbf{K}}$ and the Hamiltonian matrix becomes

$$\begin{aligned} H_{(\mathbf{k}+\mathbf{K}'),(\mathbf{k}+\mathbf{K})} &= \left\langle \exp[i(\mathbf{k}+\mathbf{K}') \cdot \mathbf{r}] \left| \hat{h}(\mathbf{r}) \right| \exp[i(\mathbf{k}+\mathbf{K}) \cdot \mathbf{r}] \right\rangle \\ &= \frac{1}{2}|\mathbf{k}+\mathbf{K}|^2 \delta_{\mathbf{K}',\mathbf{K}} + v(\mathbf{K}-\mathbf{K}') \\ &\quad + v_{\text{H}}(\mathbf{K}-\mathbf{K}') + v_{\text{xc}}(\mathbf{K}-\mathbf{K}'). \end{aligned} \tag{2.95}$$

In H the kinetic energy is diagonal and the external, Hartree, and exchange–correlation potentials are represented by their Fourier transforms.

Pseudopotentials

In calculations it is practical to split up the basis set into *core states* and *valence states*. All electron levels which are fully occupied with respect to the main quantum number n, are core states and all partially occupied levels are valence states. If atoms interact usually only the valence electrons are involved and the core electrons are inert. Therefore it is an excellent approximation to assume that the core electrons are the same in isolated atoms, molecules, and solids. This is the so called *frozen core* approximation.

A *pseudopotential* is the sum of the nuclear potential and the (frozen) core states. And it represents an effective interaction felt by the valence electrons. In this way the core states are omitted from the basis set and only the valence states are considered in the actual calculation. This concept was first proposed by Fermi [101].

If plane waves are used as basis set, the pseudopotential approximation is furthermore necessary to keep down the basis set to a manageable size. This is because the Coulomb potential in $v(\mathbf{r})$ (Eq. 2.17) is divergent at the nuclei, i.e., for $\mathbf{r} = \mathbf{R}_\alpha$. And it causes the valence wave functions to have strong oscillations close to a nucleus, because the kinetic energy increases as the potential tends to minus infinity. Alternatively it can also be seen as result of the valence states being orthogonal to the core states, which forces them to have strong oscillations (see [75]). These strong oscillations have to be represented by high–frequency (high–energy) Fourier components in the basis set. This means that a very big cutoff energy and thus a very big basis set is necessary, which would make calculations computationally expensive. However, this problem can be overcome by the use of pseudopotentials. Here the core states screen the nuclear potential such that the divergence is removed, the oscillations are reduced, and the basis set has a manageable size.

The pseudopotential concept will now be illustrated by the *Phillips–Kleinman construction* [102]. Let the core and valence eigenstates of a one–particle Hamiltonian $\hat{h}^0 = \left[-\frac{1}{2}\nabla^2 + v(\mathbf{r})\right]$ be denoted by $|\phi_c\rangle$ and $|\phi_v\rangle$, respectively. The indices c and v each represent the whole set of quantum numbers. We consider the following ansatz for $|\phi_v\rangle$, known from the method of *Orthogonalized Plane Waves* (OPW) [103, 75]

$$|\phi_v\rangle = |\phi_v^{\text{ps}}\rangle + \sum_c b_{cv} |\phi_c\rangle$$
$$\text{with: } \quad b_{cv} = -\langle \phi_c | \phi_v^{\text{ps}} \rangle. \tag{2.96}$$

The valence wave function is represented as a sum of a smooth part $|\phi_v^{\text{ps}}\rangle$, called the *pseudo wave function*, and an oscillating part, resulting from the orthogonalization of the valence state to the core states: $\langle \phi_c | \phi_v \rangle = 0$. This condition determines the expansion coefficients b_{cv}. A minor bit of algebra reveals that the pseudo wave function obeys

$$\left(-\frac{1}{2}\nabla^2 + \hat{v}^{\text{ps}}\right) |\phi_v^{\text{ps}}\rangle = \varepsilon_v |\phi_v^{\text{ps}}\rangle \tag{2.97}$$

$$\text{with: } \quad \hat{v}^{\text{ps}} = v(\mathbf{r}) + \hat{v}^{\text{nl}} \quad , \quad \hat{v}^{\text{nl}} = \sum_c (\varepsilon_v - \varepsilon_c) |\phi_c\rangle \langle \phi_c|, \tag{2.98}$$

where $|\phi_c\rangle\langle\phi_c|$ is the projection operator onto the core states, and ε_v and ε_c are the energies of the valence and core states, respectively. Relation 2.97 is a Schrödinger equation for the pseudo wave function subject to an effective potential v^{ps} – the pseudopotential. The latter is the sum of the *local* external potential $v(\mathbf{r})$ in $\hat{h}^0$ and a *non–local* potential $\hat{v}^{\text{nl}}$ (see Eq. 2.98). Non-local means that $\hat{v}^{\text{nl}}$ does not only depend on the position $\mathbf{r}$ (just like $v(\mathbf{r})$) but also on other positions that are averaged over the projection operators. The pseudopotential method has the following properties: First, it does not alter the valence energy levels, as the pseudo wave function $|\phi_v^{\text{ps}}\rangle$ in Eq. 2.97 has the same energy ε_v as the valence wave function $|\phi_v\rangle$. Second, the pseudo wave function and the valence wave function differ only close to a core region. At a certain distance away from the core, specified by the core radius (or cutoff radius) r_c, the core orbitals vanish and $|\phi_v^{\text{ps}}\rangle$ and $|\phi_v\rangle$ as well as v^{ps} and v coincide (see Eqs. 2.96 and 2.98). Third, since the energy of the valence sates ε_v is bigger than the energy of the core states ε_c, $\hat{v}^{\text{nl}}$ has a positive sign and is thus repulsive. It can therefore compensate the attraction of the nuclear Coulomb potential in v and realize the above mentioned screening. Close to a nucleus the pseudopotential v^{ps} is

2.5. BASIS SETS

therefore much weaker than the Coulomb potential in v such that the pseudo wave function $|\phi_v^{\text{ps}}\rangle$ is smooth and can easily be expanded in a plane wave basis set. This construction also explains the remarkable success of the nearly–free electron model [75] for the description of many metals and semiconductors.

In Eq. 2.98, $\hat{v}^{\text{nl}}$ depends on the eigenenergy of the electronic states ε_v ones wishes to find. This energy dependence is removed by generalizing the pseudopotentials to

$$\hat{v}^{\text{ps}} = v^{\text{loc}}(r) + \sum_{l\tau\tau'} D_{l\tau\tau'} |\beta_{l\tau}\rangle\langle\beta_{l\tau'}|. \quad (2.99)$$

Now $\hat{v}^{\text{ps}}$ is the pseudopotential of a single ion. It is spherically symmetric and therefore parameterized in spherical coordinates (r, θ, ϕ). $v^{\text{loc}}(r)$ is a local potential and the core projection operators have been replaced by generalized projection operators $|\beta_{l\tau}\rangle\langle\beta_{l\tau'}|$. There the dependence on the atomic quantum number l and a further degree of freedom τ are indicated. In practice only one or two projectors are used in the summation over τ. If only one is used the pseudopotential becomes the one of Kleinman and Bylander [104]. The potential is said to be *separable*, i.e., it is split up into a local part, representing the long–range interactions, and a non–local part for the short–range interactions. Similar to the core states, the projectors vanish outside of r_c. When these projection operators act on the pseudo wave function, it is decomposed into components with different angular momentum (quantum number l). Then each of these l-components "feels" a different non–local potential. Due to this l–projection, each component can only be orthogonalized to the lower lying core states of the same angular momentum. That is, the $3s$–component of a valence state can only be orthogonalized to the $2s$ and $1s$ states but not to the $2p$ state. This leads to problems for elements with $2p$, $3d$, or $4f$ valence states because core states of the same angular momentum do not exist. Thus the respective p, d, and f components of the pseudopotential are relatively hard and E_{cutoff} has to be big. However this problem can be overcome by the use of *ultrasoft* pseudopotentials, which will be introduced below.

In practice *ab initio* pseudopotentials are generated for each element and electronic reference configuration from suitable parameterizations that are fitted to atomic all–electron (ae) calculations such that the eigenenergies are correctly reproduced and the all-electron and pseudo wave functions and potentials coincide for $r > r_c$. To keep the plane wave cutoff energy E_{cutoff} low the pseudo wave function should also be nodeless. Good pseudopotentials are transferable between atomic, molecular, and solid systems, but they depend on the exchange–correlation functional in use. For good transferability it is necessary that the pseudopotential is *norm–conserving*. That means that the pseudo wave function must obey

$$\langle\phi_v^{\text{ps}}|\phi_v^{\text{ps}}\rangle = \langle\phi_v^{\text{ae}}|\phi_v^{\text{ae}}\rangle. \quad (2.100)$$

Since the all–electron and the pseudo wave functions are the same for $r > r_c$ by construction, condition 2.100 requires the radial part $\phi_v^{\text{ps}}(r)$ to obey for $r < r_c$

$$\int_0^{r_c} r^2 |\phi_v^{\text{ps}}(r)|^2 \, dr = \int_0^{r_c} r^2 |\phi_v^{\text{ae}}(r)|^2 \, dr. \tag{2.101}$$

This means that in the core region the *charge* of the all–electron and pseudo wave functions must be the same for every state v. Norm–conservation automatically leads to good scattering properties of the pseudopotential, i.e., the two wave functions coincide not only for the atomic eigenenergies but also for solutions with energies nearby. This then leads to good transferability.

In Troullier–Martins pseudopotentials [105] not only the norm–conservation is taken care of but also the smoothness of the pseudo wave function is maximized. This leads to a further reduction of E_{cutoff}.

In this thesis *ultrasoft* pseudopotentials by Vanderbilt were used [106]. Above we already mentioned that this method allows to generate soft pseudopotentials even for the cases of $2p$, $3d$, and $4f$ elements, that are usually problematic. Vanderbilt realized that the condition of norm–conservation requires the pseudopotential to have a certain hardness. If this constraint is dropped, the potentials can be made much softer. However, dropping norm–conservation implies that charge density in the core region is lost. But this "missing" charge density can be taken into account by redefining the *valence* charge density to be

$$\rho(\mathbf{r}) = \sum_v^{\text{occ}} \left[|\phi_v^{\text{ps}}(\mathbf{r})|^2 + \left\langle \phi_v^{\text{ps}} \left| \hat{K}^{\text{nl}}(\mathbf{r}) \right| \phi_v^{\text{ps}} \right\rangle \right]. \tag{2.102}$$

Here the first term is equivalent to Eq. 2.23 and the second term represents the "missing" valence charge density at the atomic cores and it is called the *augmentation charge*. In this way the Vanderbilt pseudopotentials are norm–conserving in a generalized sense, and it holds

$$\langle \phi_v^{\text{ps}} | \hat{S} | \phi_v^{\text{ps}} \rangle = \langle \phi_v^{\text{ae}} | \phi_v^{\text{ae}} \rangle$$

$$\text{with:} \quad \hat{S} = \hat{1} + \hat{N}^{\text{nl}} \quad , \quad \hat{N}^{\text{nl}} = \int \hat{K}^{\text{nl}}(\mathbf{r}) d\mathbf{r}, \tag{2.103}$$

where $\hat{S}$ is the overlap operator. Then the Schrödinger equation for the pseudo wave function becomes

$$\hat{h} | \phi_v^{\text{ps}} \rangle = \varepsilon_v \hat{S} | \phi_v^{\text{ps}} \rangle. \tag{2.104}$$

Here $\hat{h}$ is the DFT Hamiltonian as in Eq. 2.28, with the external potential $\hat{v}$ given by

$$\hat{v}(\mathbf{r}) = \sum_{\alpha=1}^{\Gamma} \hat{v}^{\text{ps}}(\mathbf{r} - \mathbf{R}_\alpha), \tag{2.105}$$

2.5. BASIS SETS   47

where $\hat{v}^{\text{ps}}$ is parameterized as in Eq. 2.99 and the operator $\hat{K}^{\text{nl}}(\mathbf{r})$ (in Eqs. 2.102 and 2.103) is expanded in the same set of projection operators as the non–local part of $\hat{v}^{\text{ps}}$. This formalism significantly reduces the hardness of pseudopotentials and allows to do plane wave calculations with unprecedented small cutoff energies.

For more information about the pseudopotential method see [107, 108, 109].

2.5.3 Discussion

We will briefly discuss the advantages and drawbacks of the two above described methods (basis sets) for the studies in this thesis.

The pseudopotential method is based on the frozen–core approximation and in the actual calculation only the valence states are considered. The latter are "pseudized" in the core region, i.e. the all–electron and the pseudo wave function are different. The size of the core region is defined by a fixed core radius r_c, that is chosen to be small enough to produce reliable results for each element at ambient conditions (at the equilibrium volume), but still big enough to give reasonably low plane wave cutoff energies E_{cutoff}. If such pseudopotentials were used at high pressures, where the atomic volumes are significantly smaller than the equilibrium volume, the pseudized core regions would occupy a large percentage of the available volume and might even overlap. This would certainly lead to highly incorrect results. Thus at high pressures care must be taken and pseudopotentials with small core radii r_c (and a larger E_{cutoff}) should be used. An advantage of the pseudopotential method is that interatomic forces are easily obtained, making structural optimizations fast and efficient.

LMTO is an all–electron method that accurately treats both the valence and the core electrons and is therefore predestined for the study of materials under pressure. However, the calculation of interatomic forces is relatively expensive in LMTO (at least in the LMTART program that we used [48]). This makes structural optimizations difficult.

Parts of our studies will require structural optimizations to be performed in materials at high pressures. In these cases we will combine the two methods and do the basic optimizations with the pseudopotential method and the calculations of the physical properties with the full potential LMTO method.

Chapter 3
Novel Phases of Elemental Boron

3.1 Introduction

The picture on elementary boron chemistry has changed significantly during the last ten years. The conventional paradigm that boron structures are based on B_{12} icosahedra was shifted by a so called Aufbau principle, formulated by Ihsan Boustani [22]. It is a very general structural rule that predicts the existence of quasiplanar (sheets), tubular (nanotubes), and spherical (fullerenes) boron clusters. Some of these predictions have already be confirmed, as small quasiplanar clusters and nanotubes were found in experiments [19, 21]. However, the Aufbau principle is very general and the experimental studies are not very detailed yet. Therefore questions about the precise atomic structure of boron nanotubes (BNTs) and boron sheets (BSs) remain open and theories describing their properties are needed. A general introduction to elemental boron will be given in Sec. 3.2.

Carbon nanotubes [2] are a structural paradigm for all nanotubular materials and they can be seen as tubular modifications of graphene, which can be constructed geometrically by cutting a rectangular piece out of a single graphene sheet and rolling it up to form a tube. Almost all properties of carbon nanotubes can be derived from the properties of a single graphene sheet, which means that a profound understanding of graphite is the key to understand the basic properties of carbon nanotubes. The same relation holds for broad BSs[1] and BNTs: understanding the structure and the properties of broad BSs will be crucial for our understanding of the basic properties of BNTs.

At the moment in which we carried out the studies contained in this chapter, no systematic studies of boron nanotubes and broad BS existed, and different structure

---

[1] We want to point out the distinction between the relatively small quasiplanar clusters that were found experimentally, and which are strongly influenced by finite size effects, and broad (infinite) boron sheets where these effects are unimportant. Both are called "boron sheets" in the literature.

models could be found in the literature. Our findings in sections 3.3 and 3.4, which build on on previous work [110], define a consistent picture of boron sheets and boron nanotubes, and they unify former studies on these materials into one generalized theory. Using *ab initio* structural optimization methods for solid state systems we could finally discriminate among different structure models for layered boron compounds and establish a simple model for a broad and stable BS. After a detailed description of this search process, we will analyze the properties of the most stable structure model. Then we will show how these results may be used to explain the structure, the stability, the electronic and mechanical properties of BNTs.

In section 3.5 we will turn our attention to bulk boron systems. The discovery of high–pressure superconductivity in elemental boron in the year 2001 [9] has put boron into the focus of experimental and theoretical groups worldwide. So far, the basic mechanism behind superconductivity is not understood. The main difficulty is our rudimentarily knowledge about high–pressure phases of boron. As boron tends to from multiple allotropes it is very likely that further unknown phases exist. We therefore extended the ideas behind the Boustani Aufbau principle, which was only developed for elemental clusters, to the bulk domain, and ask whether layered bulk phases, similar to graphite, may also exist for boron. Our study will try to given an answer to the following questions: What do such layered bulk structures look like? What is their stability in comparison to other bulk phases? Are these phases dynamically stable and if so, are they responsible for the high–pressure superconductivity of elemental boron?

Our results will show that novel metallic bulk phases of boron, different from the known icosahedral phases, are likely to exist at elevated pressures or even at ambient conditions, and that there are strong indications that these phases could be conventional superconductors with considerable high superconducting transition temperatures.

3.2 Fundamentals and Methods

3.2.1 Boron Chemistry

Boron is the fifth element in the periodic table. The atomic ground state configuration is $1s^2 2s^2 2p^1$. As the s-p promotion energy is only 3.58 eV (carbon 4.18 eV) boron favors sp hybridization. Similar to carbon, it forms directed covalent bonds but it is more flexible in their orientation and bond length. Typical boron-boron bond lengths are between 1.6 and 1.85 Å, typical coordination numbers are between 4 and 7.

We will see in Sec. 3.2.2 that elemental boron and boron–rich solids form structures of remarkable complexity. To understand the complex behavior of boron it is

3.2. FUNDAMENTALS AND METHODS

important to note that it falls into the class of the so called *electron deficient* elements. These are elements that have more valence orbitals than valence electrons [15]. In case of pure boron this means that the valence shell of boron has four orbitals (s, p_x, p_y, p_z) but only three electrons. So in the second row of the periodic table lithium, beryllium, and boron are electron deficient but carbon, the fourth element, is not. It has just as many valence electrons as valence orbitals and therefore it will primarily form directed covalent bonds, with coordination numbers not higher than four (three in graphite and four in diamond). For electron deficient elements Pauling postulated three empirical rules [15]:

1. The ligancy[2] of an electron deficient atom is higher than the number of valence electrons and even higher than the number of valence orbitals.

2. Electron deficient atoms cause adjacent atoms to increase their ligancies to values greater than the orbital number.

3. Typical electron deficient materials are metallic.

Let us illustrate these rules in the case of boron. The first rule is readily checked if we look at the preferred coordination of boron in its elemental bulk phases in Fig. 3.1(a), which is six, i.e., higher than the number of valence electrons and higher than the number of valence orbitals. As mentioned above, boron can occupy a spectrum of coordination numbers ranging from 4 to 7, but the most stable is always six. To verify the second rule let us notice that carbon can reach coordination numbers of five or even six if it is surrounded by electron deficient atoms (like Li or B). These structures are usually three-dimensional, e.g. CLi_6, ortho-$C_2B_{10}H_{12}$, etc. but there are also planar compounds, as shown theoretically by von Ragué Schleyer *et al.* [111, 112]. These findings may be astonishing at first glance but they are in full agreement with Pauling's second rule. The third rule seems to be violated for boron, as all known bulk phases are semiconducting (see Sec. 3.2.2). But, as we will show in Secs. 3.3, 3.4 and 3.5, modifications related to planar or quasiplanar boron are always metallic.

In materials where boron is *not* the dominating atomic species it usually acts as electron acceptor and forms sp^2 or sp^3 bonds. An example are the metal diborides of composition MB_2 where, in a local chemical picture, the metal atom (M) transfers its valence electrons to boron, which then becomes isoelectronic to carbon in graphite and forms a planar or puckered sp^2 honeycomb lattice.

If the boron content of a compound rises further, like in the higher metal borides MB_4, MB_6, or MB_{12}, the electron transfer from the metal atoms will be insufficient to overcome the electron deficiency of all boron atoms if we think in terms of usual two-center sp bonding, i.e. two atoms sharing two electrons. But an efficient way of

---
[2]The term *ligancy* means the coordination number.

phase	AB	N_{atom}	SG	LP	ρ	color
α-rhomb.	R-12	12	$R\bar{3}m$ (166)	$a = 5.064$ Å $\alpha = 58.10°$ [113]	2.46	red and translucent
β-rhomb.	R-105	106.7	$R\bar{3}m$ (166)	$a = 10.14$ Å $\alpha = 65.21°$ [114]	2.33	shiny gray
α-tetra.	T-50	50	$P4_2/nnm$ (134)	$a = 8.75$ Å $c = 5.06$ Å [115]	2.31	black
β-tetra.	T-192	189.9	$P4_3$ (78)	$a = 10.14$ Å $c = 14.17$ Å [116]	2.36	red

Table 3.1: The most common phases of elemental boron and their properties. The T-50 phase is not generally accepted as real allotrope of pure boron (see text). AB is a traditional abbreviation according to [35], N_{atom} is the mean number of atoms per unit cell, SG is the space group, the number in parenthesis is the space group number according to Ref. [117], and LP are the lattice parameters, and ρ is the density in g/cm^3.

sharing the electrons in that situation are *three-center two-electron bonds*. A three–center bond derives from a situation in which three atomic orbitals, each associated with different atoms, can be combined to give a single lowest energy molecular orbital. Two electrons are then shared by three atoms and the bonding is mediated by charge density that is accumulated in the center of a *triangle* formed by three atoms.[3] This picture can be generalized for n–center bonding. The associated bonds are called *multi–center bonds* [16, 17, 118]. Generally, the bonding in elemental and boron-rich materials is a complex "mixture" of two-center and three–center bonds. Both are of covalent type but three–center bonds are less localized.

In elemental boron and boron–rich materials the boron atoms form three–dimensional networks of polyhedra (i.e., B_{12} cubo-octahedra in MB_{12}, octahedra in MB_6, icosahedra in elemental boron) or fragments of polyhedra in which triangular faces prevail[4] and which are stabilized by foreign atoms sitting at interstitial sites. The triangular B-B-B faces are a consequence of three–center bonds, which are a crucial concept to understand the bonding in these materials.

All elemental bulk modifications are based on a three–dimensional framework of slightly distorted B_{12} icosahedra. A regular icosahedron is a polyhedron with 12 vertices, 20 equilateral triangular faces, and 30 edges (see Fig. 3.1(b)). The icosahedra are strongly bound to each other, and an "inverse umbrella" coordination with a pentagonal pyramid as coordination polyhedron (see Fig. 3.1(a)) is the preferred atomic coordination within the bulk. Furthermore, any B_{12} icosahedral network is always

---

[3] In a two–center bond the charge density is accumulated along a *line* connecting two atoms.
[4] A polyhedron whose faces are only equilateral triangles is called a *deltahedron*.

3.2. FUNDAMENTALS AND METHODS

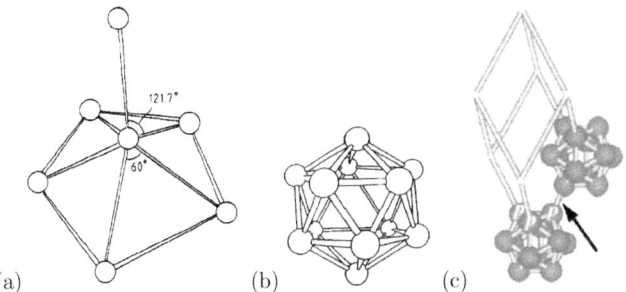

Figure 3.1: (a) The preferred "inverse umbrella" coordination of boron in bulk phases [120]. (b) A B_{12} icosahedron [26]. (c) The crystal structure of α-rhombohedral boron; the arrow points at the inter-icosahedral bond, indicating that this crystal is not a molecular solid [121].

stabilized by the introduction of foreign atoms, even in trace amounts. They either complete the coordination of boron or provide electrons to stabilize the electronic structure. Because of this strong affinity of boron to other elements it is difficult to obtain it in elemental forms.

3.2.2 Elemental Bulk Phases

Elemental boron does not exist in nature. After an extraction from minerals like borax ($Na_2B_4O_7 \cdot 10\ H_2O$ or $Na_2B_4O_5(OH)_4 \cdot 8\ H_2O$) or ulexite ($NaCaB_5O_9 \cdot 8\ H_2O$) it can occur in different forms.

Although glassy and amorphous states of elemental boron were known before, the first crystalline phase was reported in 1943 by Laubengayer et al. [119]. At least 15 further phases were reported until the mid 1970s [35]. But today only three of them are generally accepted as true polymorphs: the α-rhombohedral, β-rhombohedral, and β-tetragonal phase. The other presumed modifications are likely to be boron-rich phases instead of real elemental ones. Table 3.1 lists some of the basic properties of the most common phases.

The simplest phase is the α-rhombohedral one (R-12), where slightly distorted boron icosahedra are centered on the vertices of a rhombohedral unit cell (see Fig.3.1(c)).[5] The icosahedra are bound to each other via two and three-center bonds.

The first elemental phase that was reported is α-tetragonal (T-50) boron [119].

---
[5]The rhombohedral angle of the R-12 phase is $\alpha_{rh} = 58.1°$ (see Tab. 3.1). Note that an angle of 60° would be equivalent to a fcc lattice.

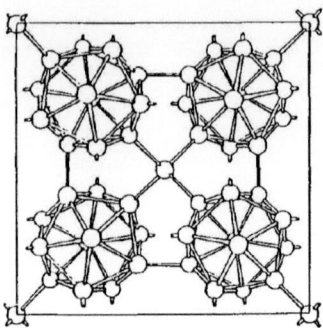

Figure 3.2: The unit cell of α-tetragonal boron contains 50 atoms [122].

It was also discussed by Linus Pauling is in famous book "The nature of the chemical bond" [15]. The 50 atoms unit cell is shown in Fig. 3.2. It contains 4 icosahedra and 2 additional 4-fold coordinates atoms at the corners and center of the unit cell. The icosahedra are strongly linked to each other via 2-center bonds only. Today there are strong doubts about the existence of T-50 boron. According to the careful experimental work of Amberger *et al.* [127, 128] the B_{50} crystals reported before were $B_{50}C_2$ or $B_{50}N_2$, i.e., a boron network stabilized by small amounts of foreign atoms. This is also in agreement with theoretical calculations [129, 120] that show that T-50 boron has too few electrons to reach a stable configuration.

As shown in Fig. 3.3(a) the β-rhombohedral modification (R-105) has icosahedra at the vertices and also half-way between two vertices of the rhombohedral unit cell. The interstitial space is filled with a B_{28}-B-B_{28} chain, where the B_{28} unit is formed by three fused icosahedra (see Fig. 3.3(b)). The structure of R-105 boron can alternatively be based on the B_{84} superunit, shown in Figure 3.3(c). It consists of a central B_{12} icosahedron where each of the 12 atoms is radially connected to a B_6 unit via the preferred coordination (see Fig. 3.1(a)): $B_{84} = B_{12} + 12\ B_6$. The outer atoms of this arrangement form a bucky ball (fullerene), a polyhedron consisting of 12 pentagons and 20 hexagons,[6] whose pentagons are already part of neighboring icosahedra (see Fig. 3.3(e)). These B_{84} units are centered on the vertices of the rhombohedral unit cell (see Fig. 3.3(d)). Now three pentagonal pyramidal B_6 units of the above mentioned interstitial B_{28} unit are already part of the B_{84} superunit, and therefore the interstitial chain is reduced to B_{10}-B-B_{10} in this second representation.

In β-tetragonal boron (T-192) the icosahedra are arranged in "stacked" linear

---

[6]Note that an *ideal* bucky ball has icosahedral symmetry just like the enclosed icosahedron. However, the actual structures are less symmetric (space group $R\bar{3}m$).

3.2. FUNDAMENTALS AND METHODS

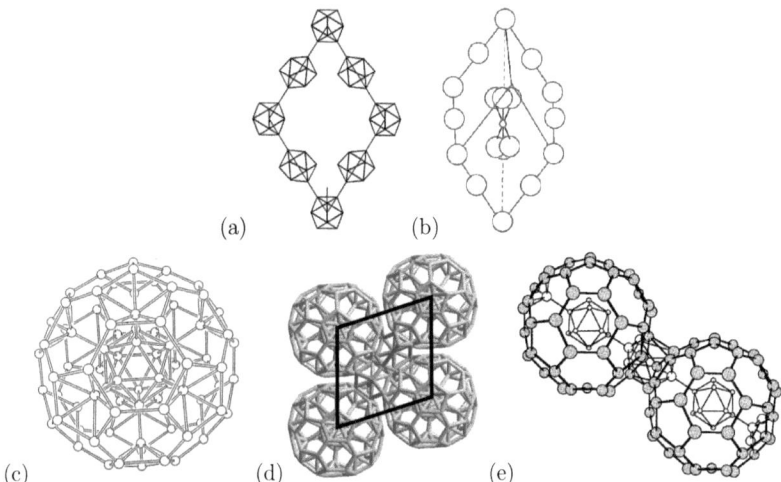

Figure 3.3: The structure of β-rhombohedral boron. (a) The distribution of icosahedra on a face of the rhombohedral unit cell [123]. (b) The positions of icosahedra (big white balls) within the unit cell (not all icosahedra on the faces of the unit cell are shown). Along the long diagonal (dashed line) a B_{28}-B-B_{28} arrangement filles the interstitial. The B_{28} subunit is formed by three fused icosahedra [124]. (c) A B_{84} superunit [125]. (d) The B_{84} units are centered on the vertices of the rhombohedral unit cell and the interstitial is filled with a B_{10}-B-B_{10} arrangement [39]. (e) The connection of neighboring B_{84} units and neighboring icosahedra along the edge of the unit cell [126]. In (d) and (e) many internal bonds have been omitted for clarity.

chains as shown in Fig. 3.4. This B_{12}-skeleton leaves huge interstitial voids which are filled with four B_{21} units (not shown in the figure). They are built from two icosahedra sharing a common triangular face. In total the unit cell contains eight B_{12}, four B_{21} units, and 16 interstitial B sites. All of them are strongly bound to each other via two-center bonds.

The above phases (except R-12) are characterized by some degree of intrinsic disorder: in Tab. 3.1 the average number of atoms per unit cell of the β-crystalline modifications is non-integer. This stems from a few *partially occupied* atomic sites. These sites are occupied by boron atoms not in all but only in some unit cells of the crystal. As shown by Slack *et al.* [114], the occupancies of these sites depend only weakly on the preparation method and are indeed intrinsic to the phase. These partially occupied sites are mainly interstitial atoms that do not belong to the B_{12}-

skeleton of the structures. Electronic structure calculations of the R-105 and T-50 phases using 105 and 50 atoms per unit cell, respectively, show that the structures are short of electrons to completely fill the conduction band [120]. Thus these phases are stabilized by electron doping via foreign atoms or partial occupied sites.

General Properties

All known bulk modifications of boron are extremely hard semiconductors. The Mohs hardness[7] of the R-105 modification is 9.3 [132], the one of diamond is 10. Boron-rich solids are generally among the hardest materials known. The hardness is probably mediated by the extremely strong three-dimensional framework of B_{12} icosahedra. All phases have very high melting temperatures ($T_m = 2335°C \pm 35°C$ for R-105 boron [10]), a low density (see Tab. 3.1),[8] and a small reactivity at room temperature. Furthermore, due to the big neutron scattering cross section of the ^{10}B isotope, boron is used as neutron absorber in nuclear reactions and in *neutron capture therapy*[9], an anti–cancer therapy.

The thermodynamic properties of boron are highly unknown. Neither there is an experimentally determined phase diagram nor is the ground state structure known. This can be attributed to the complicated polymorphism of boron and the difficulties in preparing pure boron. It is experimentally known that R-12 boron is thermally unstable above 1200°C and that above 1500°C R-105 is stable up to the melting point T_m. Between 1000 and 1500°C different crystalline forms can be grown.

In a series of theoretical papers Masago, Shirai, and Katayama-Yoshida [124, 133, 134, 10] studied the relative stability of α- and β-rhombohedral boron. They considered the effects of disorder in R-105 boron, phonons, and thermal volume expansion and found that α-boron is the thermodynamically stable phase at low temperatures and high-pressures. Very recently, van Setten et al. [135] and Widom et al. [?] theoretically refined the crystal structure of R-105. They employed the structural degrees of freedom that are connected with the partially occupied sites. Their "improved" R-105 phases are lower in free energy than R-12 and appear to be the thermodynamical ground state at low temperatures and ambient pressure. The contradictory results of the different authors show that there is demand for further studies.

---

[7]The Mohs scale of mineral hardness characterizes the scratch resistance of various minerals through the ability of a harder material to scratch a softer material [130, 131].

[8]Due to its low density and high melting temperature boron is used as thermal protection shield for spaceships and in nuclear fusion experiments.

[9]In boron neutron capture therapy the patient is given an infusion that contains boron and which collects in tumor cells. The patient then receives radiation therapy with neutrons. The tumor cells are killed by the thermal energy that is released in the nuclear reaction $^{10}B + n \rightarrow {}^{11}B^* \rightarrow {}^7Li + {}^4He$.

3.2. FUNDAMENTALS AND METHODS

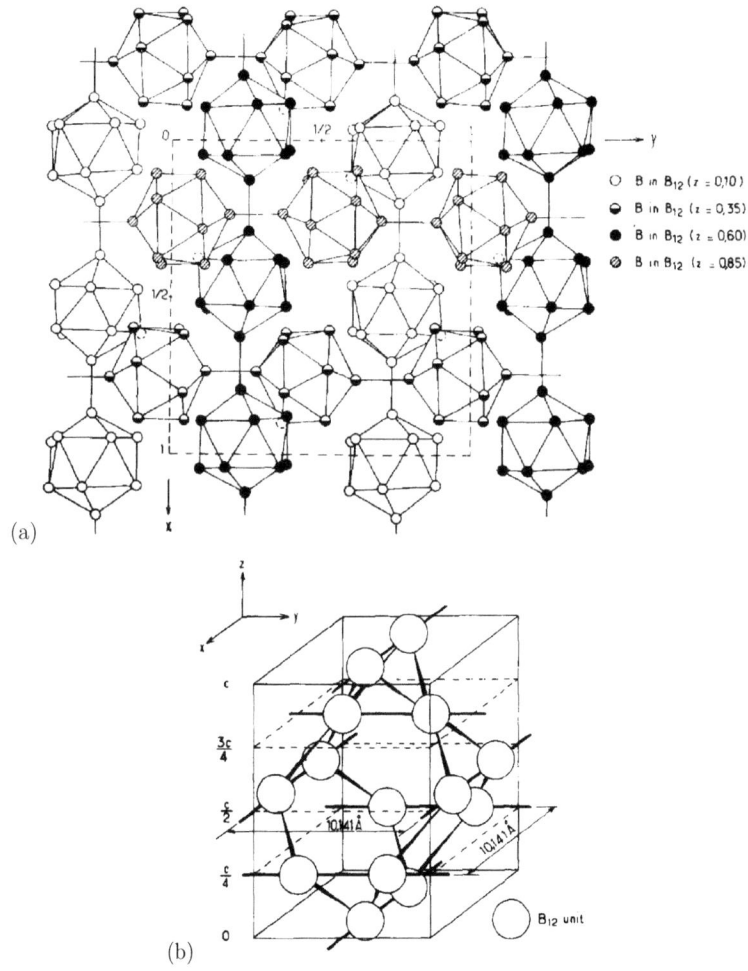

Figure 3.4: The arrangement of icosahedra within the unit cell of β-tetragonal boron [116]. (a) shows a xy projection of the unit cell (dashed-lined box) and (b) a perspective view. The boron skeleton is an arrangement of parallel chains of B_{12} units located in xy planes. Each plane is separated from the other by $c/4$.

It is theoretically and experimentally established that R-12 is stiffer than R-105 (the bulk modulus is slightly bigger, i.e., $\sim$ 220 GPa for R-12 and $\sim$ 200 GPa for R-105). This can be explained [124] by the fact that R-12 is denser than R-105 (see Table 3.1). The high density of R-12 also explains why it appears to be stable at high pressure. It that sense R-12 boron can be called *mechanically stable*, while R-105 is *thermodynamically stable* [134].

In α-rhombohedral boron (and also in the other phases) the bonds within the icosahedra (intraicosahedral) are longer than the ones between them (intericosahedral); 1.75-1.80 Å vs. 1.67 Å, respectively [113].[10] The question, which of these bonds is the stronger one, is a long lasting debate among boron researchers. Following a general rule of chemistry, the intericosahedral bonds should be stronger because they are shorter. Intuitively one would further expect the intraicosahedral bonds, which are mainly of three-center character, to be weaker than the intericosahedral ones, which are common σ bonds of two-center character. Thus R-12 boron seems to have stronger bonds between the icosahedral units than within them. Therefore R-12 boron was entitled an *inverted-molecular* solid [136]. This picture is probably right, as shown recently by works of Nelmes *et al.* [137] and Shirai *et al.* [124], but it is also contested by Fujimori *et al.* [121] and Lazzari *et al.* [138].

Due to the slight distortion of the icosahedra, and the resulting spectrum of intraicosahedral bond lengths, the charge density within an icosahedron is distributed asymmetrically. Garcia and Cohen have attributed the existence of ionicity to asymmetries of the charge density distribution [139]. According to He *et al.* [140] this viewpoint is indeed true for α-rhombohedral boron, which is the first material where ionicity is found in chemical bonds between atoms of the same type.

Eremets *et al.* [9] performed electrical conductivity measurements under pressures up to 250 GPa and discovered that R-105 boron transforms from a nonmetal to a *superconductor* at $P = 160$ GPa, with $T_c = 5$ K at 175 GPa and 11.2 K at 250 GPa. They also observed steps in the conductivity at 30, 110, and 170 GPa, possibly indicating phase transitions (or measurement problems). The superconductivity is not theoretically understood, so far. This is primarily due to almost unknown high-pressure behavior of boron.

3.2.3 Clusters, Nanostructures, and the Aufbau Principle

Until the mid–1990s elemental boron was thought of only in terms of the icosahedral bulk structures. Although the chemistry of boron was known to be complicated, little attention had been payed on pure boron materials and very little was known about boron clusters. This situation was changed by Boustani *et al.* Inspired by the work of Andersen *et al.* [141, 142], who experimentally studied small boron clusters, they

---

[10]In this discussion we exclude the intericosahedral three-center bonds in α-rhombohedral boron.

3.2. FUNDAMENTALS AND METHODS

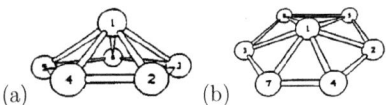

Figure 3.5: The two building blocks of the Aufbau principle: (a) the pentagonal B_6 pyramid, and (b) the hexagonal B_7 pyramid [22].

started in 1994 [143] to systematically study boron structures, beginning from the smallest cluster up to an infinite number of boron atoms [144, 145]. This theoretical work led to the development of a so called *Aufbau principle* [22] – an empirical rule that predicts the morphology of stable boron clusters.

The Aufbau Principle

The Aufbau principle states that stable boron clusters can be constructed from two basic units only: a pentagonal pyramidal B_6 and a hexagonal pyramidal B_7 unit, which are allowed to interpenetrate. These units are shown in Fig. 3.5.

Combinations of these elements lead to four topological classes of stable boron clusters: quasiplanar [23], tubular [24, 25], convex and spherical [26, 33]. Furthermore, a quasicrystalline modification based on α-rhombohedral boron was predicted by studying elliptical clusters in Ref. [146, 147]. Quasiplanar, convex, and tubular clusters are formed by combinations of hexagonal pyramidal B_7 units, as illustrated in Fig. 3.6. Combinations of B_6 and B_7 elements lead to convex and spherical clusters (boron fullerenes), as shown in Fig. 3.7.[11] Usually, those structures are not atomically smooth, but their surfaces are puckered (dimpled).

At the time of their prediction none of these structures were known in nature. Nevertheless, extensive numerical studies of many groups (see [154] and references therein) supported the findings of Boustani *et al.* In 2003 and 2004 the existence of quasiplanar boron clusters [19] and boron nanotubes [21] was experimentally confirmed. Beyond that new materials in the form of boron nanorods/nanowires [155, 156, 157], nanobelts/nanoribbons [158, 159] were found by experiment, thus opening the gates for a new field of boron based nanomaterials.

Chemistry and Stability

The basic chemistry behind the Aufbau principle is in full agreement with the general discussion in Sec. 3.2.1. The B_6 unit is a simple realization of the preferred bulk

---

[11] Closed tubular and spherical boron hydride (borane) clusters were independently studied also by Lipscomb *et al.* [148, 149, 150, 151], Jemmis *et al.* [152], and others [153]. But none of them considered elemental boron.

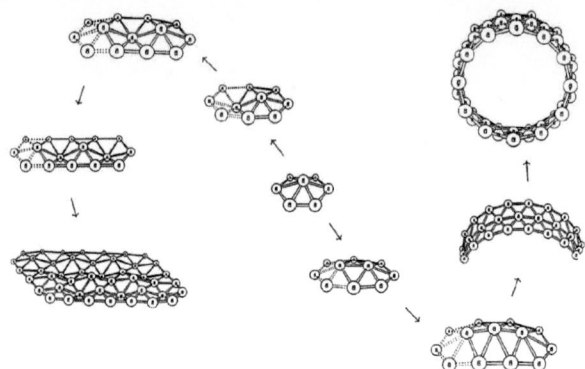

Figure 3.6: The Aufbau principle employing only B_7 units generates (left) quasiplanar and (right) convex layers that eventually form nanotubes [22].

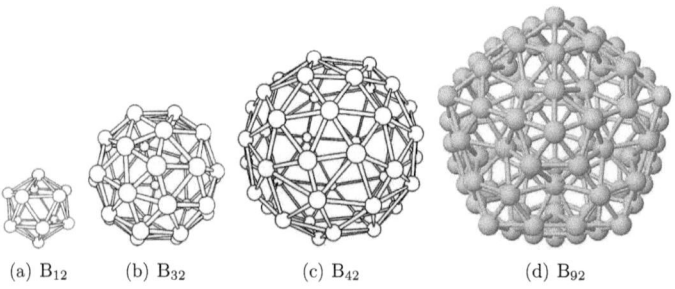

(a) B_{12} (b) B_{32} (c) B_{42} (d) B_{92}

Figure 3.7: Spherical boron clusters [26, 33].

3.2. FUNDAMENTALS AND METHODS

coordination, shown in Fig. 3.1(a). The Aufbau principle itself is a realization of the chemical observation that boron-boron bonds are primarily of three-center character, forming B-B-B triangles, and that the coordination number ideally lies between five and seven.

The B_{12} cluster in Fig. 3.7(a) is built from 12 inter-penetrating pentagonal pyramidal B_6 units. It fulfills the Aufbau principle and therefore all icosahedral bulk structures do so. But it is known since a long time that a neutral B_{12} cluster alone is unstable [129, 17]; in the bulk however the B_{12} units are stabilized by intericosahedral bonds. For *small* boron clusters (including B_{12}) it turns out that quasiplanar morphologies are the most stable configurations [22, 26]. The cluster electrons can be divided into the ones that mediate the in-plane bonding of the cluster skeleton, and delocalized π-electrons that stabilize the quasiplanar shape [23].[12] Chemists showed that the concept of aromaticity can be applied to these clusters which explains their unusual stability [160, 19, 161, 154].

From the knowledge that small boron clusters are quasiplanar one can immediately conclude that boron nanotubes and boron fullerenes should exist, because a growing quasiplanar boron cluster tends to remove dangling bonds by forming closed tubular or polyhedral modifications. This is completely analogous to carbon, where fullerenes and nanotubes only grow out of the cluster phase and are a compromise between the energy cost for bending the planar cluster and the energy gain for removing dangling bonds. Therefore, with growing cluster size tubular and spherical clusters will become more stable than quasiplanar ones. And indeed, Kiran *et al.* [20] reported the B_{20} cluster to be the planar-to-tubular transition point, and Szwacki *et al.* [33] theoretically determined the B_{80} cluster to be the smallest spherical cluster (boron fullerene).

This shows that the picture on elementary boron chemistry has changed significantly during the last ten years. The conventional paradigm that boron structures are based on B_{12} icosahedra was expanded by the Aufbau principle. And an interesting new field of research ranging from cluster chemistry, nanoscience, solid state physics to basic research on superconductivity has emerged and is going to expand further.

3.2.4 Computational Details

As showed in the last sections, elemental boron has a complicated and rather versatile chemistry and many aspects are not understood, so far. Therefore the only reliable theoretical tools, which may allow for a proper description of boron chemistry, are

---

[12]In fact, the main difference between the much-studied boron hydrides (boranes) and elemental boron clusters is the occurrence of delocalized π-electrons. In boranes these electrons are "pinned" by the terminating hydrogen bonds leading to the preference of polyhedral (deltahedral) morphologies instead of (quasi)planar ones.

first principles calculations. The details of the computations that are presented in this chapter will be described now.

In order to carry out structural optimizations of boron sheets (BSs), boron nanotubes (BNTs), and layered bulk structures in Secs. 3.3, 3.4, and 3.5 we used the VASP package, version 4.4.6 [162, 163]. The latter is an *ab initio* code based on density functional theory [60] – using plane wave basis sets to model solid materials, surfaces, or clusters [164]. For the simulations of BSs and BNTs, the electronic correlations were treated within the local–density approximation (LDA) using the Perdew–Zunger form of the Ceperley–Alder exchange–correlation functional [62, 81]. For the bulk simulations in Sec. 3.5 we used the generalized gradient approximation (GGA) functional "Perdew-Wang 91" [165]. The ionic cores of the system were approximated by ultrasoft pseudopotentials [106] supplied by Kresse and Hafner [166]. The k–space integrations were carried out using the method of Methfessel and Paxton [167] in first order, with a smearing width of 0.3 eV. The cutoff energy for the expansion of the electronic wave functions in terms of plane waves was 257.1 eV for the relaxation runs, and 321.4 eV for a final static calculation of the total energy. The total energies were converged such that changes in the total energies were less than 10^{-4} eV.

With the help of the VASP program, one can determine interatomic forces, which may be used to relax the different degrees of freedom for a given unit cell. Eventually one will detect some atomic configurations, which correspond to (local) minima on the total energy landscape. In order to carry out those extensive structure optimizations in a effective way, we employed a conjugate gradient algorithm [164], and we allowed all of the atomic coordinates to relax, as well as all but one lattice parameter. This rigid lattice parameter fixes the interlayer separation for BS and the intertubular distance for BNTs at 6.4 Å, which effectively makes them stand-alone objects. For the bulk systems we optimized *all* lattice parameters.

The cohesive energies given in Tables 3.3 and 3.4 were calculated from

$$E_{\mathrm{coh}} = E_{\mathrm{bind}}/n. \tag{3.1}$$

E_{bind} is the the atomic binding energy per unit cell and n is the number of atoms per unit cell. Therefore in our definition E_{coh} will be a positive number.

For band structures and the analysis of Fermi surfaces of BSs in Secs. 3.3.2 and 3.3.3 we used the STUTTGART TB-LMTO-ASA package, version 4.7, which is a density functional theory [60] based code using short range [98] linearized muffin-tin orbitals [92] within the atomic spheres approximation (ASA). It allows static calculations of the electronic properties for periodic systems. We used the non-spin polarized LDA exchange-correlation functional of von Barth and Hedin [61] and a k-mesh of 30 x 30 x 3.

Charge densities or electron localization functions (ELFs) [168] were calculated either with VASP or TB-LMTO-ASA. Isocontour plots which are frequently used

3.2. FUNDAMENTALS AND METHODS

Phase	k–mesh	V (Å^3/atom)	$E_{\text{cut}}^{\text{LMTART}}$ (Ry)
BSs	6 x 6 x 4	–	–
zigzag BNTs	1 x 1 x 15	–	–
armchair BNTs	1 x 1 x 10	–	–
fcc	20 x 20 x 20	4.00 – 8.00	284 – 179
R–12	8 x 8 x 8	4.00 – 8.20	470 – 339
α–Ga	12 x 12 x 12	3.75 – 6.75	543 – 360
Immm	8 x 12 x 4	4.00 – 8.50	534 – 375
Fmmm	10 x 10 x 10	4.40 – 8.25	414 – 295

Table 3.2: Sizes of k–point meshes used in VASP and LMTART and plane wave cutoff energies in LMTART. The different models for boron sheets (BSs) are discussed in Sec. 3.3, free–standing zigzag and armchair boron nanotubes (BNTs) in Sec. 3.4, all other phases are bulk structures and are considered in Sec. 3.5. The cutoff energy in LMTART affects the representation of charge densities and potentials in the interstitial region, however it has no influence on the basis set. As $E_{\text{cut}}^{\text{LMTART}}$ depends on the atomic volume V the minimal and maximal volumes that were considered are given in the third column and the corresponding cutoff energies in the last column.

in this chapter are qualitative in character, and therefore the results of both codes are essentially the same.

The bulk structures discussed in Sec. 3.5 were reexamined with the full-potential LMTO (FP–LMTO) density-functional code LMTART developed by S. Y. Savrasov [48, 92], version 6.8. We employed a triple–κ sp basis set[13] and represented the charge densities and the potentials by spherical harmonics up to $l = 6$ inside non-overlapping muffin–tin spheres and by plane waves between the spheres. The plane wave cutoff energies, which are determined automatically in LMTART 6.8, are given in Tab. 3.2. The k–space integration was performed with the improved tetrahedron method [169, ?] and electronic correlations were treated with the "Perdew-Wang 96" GGA exchange-correlation functional [170]. The total energies were converged until changes were less than 10^{-6} Ry.

For the calculation of total energies and atomic forces (VASP and LMTART) the sizes of the k–point meshes were individually converged for different systems, such that changes in the total energy were reduced to less than 3 meV/atom; they are given in Tab. 3.2. In the course of a structural optimization all interatomic forces were reduced to less than 0.04 eV/Å.

The optimizations of the bulk systems were first done with VASP and then repeated with LMTART. We did that to take advantage of the automated optimization

---
[13] The inclusion of $3d$ states into the basis set had no influence on the band structure. The orbital characters of the tails of the basis function on neighboring atoms are taken into account up to $l = 6$.

methods implemented in VASP and because all–electron methods like FP–LMTO are more accurate at high pressures (small atomic volumes) than standard pseudopotential implementations. The structural differences between VASP and LMTART for the lattice parameters were less than 1%; internal parameters (atomic positions) were corrected by about 2%. Since the overall agreement between the two methods is good in Sec. 3.5, we mainly present the results obtained with LMTART which are: structural details, the $T = 0$ phase diagram, the electronic and phononic structure, and the electron-phonon coupling.

The phonon frequencies and the electron–phonon linewidths were calculated via linear response/density functional perturbation theory (DFPT) as implemented in LMTART [49] using the same numerical parameters for the electronic system as described above. The k–point meshes for the representation of the induced charge densities were the same as the ones used for the total energy calculations (see Tab. 3.2). This ensured that the phonon frequencies were converged within 2 meV. The sizes of the phonon q–point meshes will be discussed later on. The k–meshes, representing the electronic structure of the non–perturbed system in the linear response calculations, were 48^3, $48 \times 72 \times 24$, and 40^3 for boron in the α–Ga structure, Immm, and Fmmm, respectively. Integrations over the Brillouin zone in k– and q–space were again carried out using the tetrahedron method [169, ?]. The induced charge densities were converged until the root mean square changes were less than 10^{-6}.

For data processing and visualization the programs MATLAB, MAPLE, XM-GRACE, XCRYSDEN [171], and OPENOFFICE were used. The symmetries of the crystal structures as well as the symmetries of the phonon modes were determined by the ISOTROPY package [172].

3.3 Broad Sheets

In this section we will employ first principles simulations to find structure models of broad boron sheets (BSs). A broad sheet is the boron analog of a single graphene sheet. Furthermore, it is the precursor of boron nanotubes, similar to graphene being the precursor of carbon nanotubes (CNTs). Up to now, a broad BS could not be found in experiment; so far only the existence of small quasiplanar boron clusters [19] and boron nanotubes (BNTs) [21] is experimentally supported. But the missing link between the two systems must be the broad BS, which is quasi implied by the existence of BNTs (see Sec. 3.2.3). A detailed knowledge of the properties of a broad BS will be essential to understand the properties of BNTs. After a detailed description of the search process, we will analyze the properties of the most stable structure model.

At the moment in which we carried out the studies contained in this chapter, we became aware of an interesting work by Evans et al., [173] who consider three

3.3. BROAD SHEETS

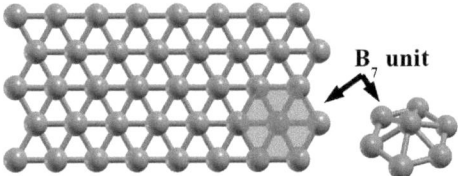

Figure 3.8: Top view of a quasiplanar boron sheet. In a planar projection the atoms form an almost perfect triangular lattice. The basic structural unit is a hexagonal pyramidal B_7 cluster, as suggested by the Aufbau principle [22] (see text).

BS models and five BNTs of small tube radii, and the work of Cabria *et al.* [174] who study two BS models and three BNTs. Although our results are certainly based on a more extensive search for stable BS and BNTs, our findings for the stable BS are, from a structural and energetic point of view, in excellent agreement with these authors. Thus the present structure model could independently be confirmed by three different groups. However, Lau *et al.* [175] have recently reported about structures for BS and BNTs, which are very different from the structure models of Evans *et al.* and Cabria *et al.*, but the present study is in clear favor of the latter.

3.3.1 Finding Structure Models

Following the Aufbau principle (see Sec. 3.2.3) a BS is basically a quasiplanar arrangement of hexagonal pyramidal B_7 units. A planar projection of such a system will always form a nearly triangular lattice (see Fig. 3.8). However, the out of plane modulation (i.e., the puckering) remains unspecified by the Aufbau principle. The latter has to be determined using *ab initio* structural optimizations, after setting up a suitable supercell that will allow for a systematic generation of various periodic puckering schemes.

The versatile chemistry of boron is reflected by a complicated energy landscape, which is full of local minima. Therefore the standard optimization techniques like the conjugate gradients method used in this study are most likely to find local minima, rather than global minima. Therefore we examined the energy landscape quite carefully by performing *many* optimization runs, which started from quite diverse initial configurations.

The basic puckering schemes were taken from the structures of B_{22}, B_{32}, and B_{46} clusters, which are described in Ref. [23]. We repeated the puckering periodically in a triangular supercell containing 16 atoms (see Figs. 3.9(b)-3.9(d)), and optimized

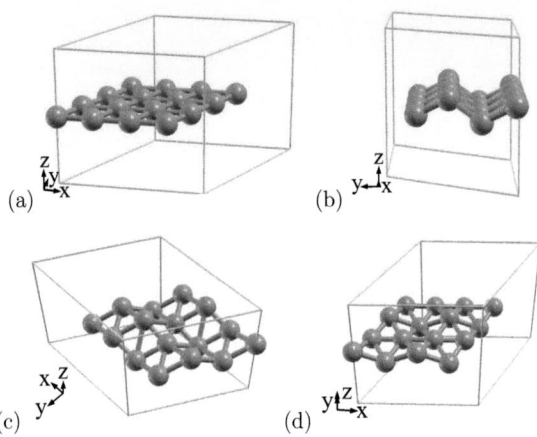

Figure 3.9: Different structure models for broad boron sheets. Each supercell (thin lines) contains 16 atoms. (a) A planar sheet is metastable. (b) Up and down puckering seems to be the most stable modulation. Structures (c) and (d) are unstable. Models (b), (c), and (d) are periodic repetitions of structural motives taken from B_{22}, B_{46}, and B_{32} clusters, described in Ref. [23].

the resulting structures.[14] For the sake of comparison we also examined a planar BS (see Fig. 3.9(a)). The planar boron sheet (a) occupies a local minimum on the energy landscape with a cohesive energy of 6.76 eV/atom. However, small out-of-plane elongations of individual atoms immediately cause a puckering of the BS. This was confirmed by shifting one atom 0.1, 0.2, and 0.4 Å out of plane and re-optimizing the resulting structures. Thus model (a) turns out to be metastable (as also pointed out in Refs. [173] and [174]); any thermal vibration would lead to a permanent deformation of a planar boron sheet. Models (c) and (d) are completely unstable, and they both relax to structure (b). In order to scan the energy landscape for other candidate structures we took sheet (a) and shifted each of the 16 atoms out of the plane, employing a random elongation Δz between $+0.4$ and -0.4 Å. Those structures were re-optimized as before. It turns out that 8 out of 11 optimizations led to model (b), while the remaining three runs resulted in a metastable kinked structure with a cohesive energy of 6.86 eV/atom (not shown).

The fact that models (c) and (d) as well as 8 out of 11 randomly puckered sheets would relax to model (b) means that structure (b) defines a rather pronounced

---

[14]The initial in-plane boron-boron distance was 1.6 Å.

3.3. BROAD SHEETS

Sheet	(a) Planar	(b) Puckered
Lattice type	Triangular (2D)	Orthorhombic (3D)
Lattice param. (Å)	$A = 1.69$	$A = 2.82$
		$B = 1.60$
		$C = $ arbitrary
Primitive vectors	$\boldsymbol{a}_1 = A(\frac{\sqrt{3}}{2}, \frac{1}{2})$	$\boldsymbol{a}_1 = A(1, 0, 0)$
	$\boldsymbol{a}_2 = A(\frac{\sqrt{3}}{2}, -\frac{1}{2})$	$\boldsymbol{a}_2 = B(0, 1, 0)$
		$\boldsymbol{a}_3 = C(0, 0, 1)$
Atoms/unit cell	1	2
Atomic pos. (Å)	$\boldsymbol{R}_1 = (0, 0)$	$\boldsymbol{R}_1 = (0, 0, 0)$
		$\boldsymbol{R}_2 = (\frac{1}{2}A, \frac{1}{2}B, 0.82)$
Bond lengths (Å)	$a_{\text{B-B}} = 1.69$	$a_{\text{B-B}}^{\sigma} = 1.60$
		$a_{\text{B-B}}^{\text{diagonal}} = 1.82$
E_{coh} (eV)	6.76	6.94
Elastic modulus	$C_x = C_y = 0.75$	$C_x = 0.42$
(TPa)		$C_y = 0.87$

Table 3.3: Detailed LDA description of the optimized lattice structures of the planar (a) and puckered (b) boron sheets (see Figs. 3.9 and 3.11), their cohesive energies E_{coh} (Eq. (3.1)), elastic moduli $C_x = C_{11}$ and $C_y = C_{22}$ obtained after stretching a sheet along the Cartesian x or y direction (Eqs. (3.2) and (3.3)), and bond lengths ($a_{\text{B-B}}^{\text{diagonal}} = a_{\text{B-B}}^{1-2}$ is the bond length between atom 1 and 2, and $a_{\text{B-B}}^{\sigma} = a_{\text{B-B}}^{1-1} = a_{\text{B-B}}^{2-2} = B$ is the bond length between two equivalent atoms in different unit cells).

minimum on the energy landscape. The high structural stability of model (b) is confirmed by its high cohesive energy of 6.94 eV/atom, which is the highest cohesive energy of all BSs that we found. We thus conclude that the most suitable structure model for a broad BS will be (b), being 0.18 eV/atom more stable (0.21 and 0.26 eV/atom in Refs. [174] and [173], respectively) than an unrealistic planar BS. The puckering itself seems to be an important mechanism to stabilize the BS [174], which will be examined in more detail in Sec. 3.3.3.

In order determine the lattice structures of (a) and (b) we performed LDA calculations, where we would fix the unit cell of each system for a series of Cartesian lattice constants A or B, whereas all of the internal (atomic) degrees of freedom were allowed to relax. The resulting total energies for a given set of lattice constants were fitted to polynomial curves $E(A)$ and $E(B)$, from which we determined the equilibrium properties of the systems. The results are summarized in Table 3.3. The elastic constants $C_x = C_{11}$ and $C_y = C_{22}$ may be interpreted as a first approximation to a macroscopic Young's modulus. They were calculated as follows:

$$C_x = \frac{A_0}{Bh}\left(\frac{\partial^2 E(A)}{\partial A^2}\right)_{A_0}, \quad (3.2)$$

$$C_y = \frac{B_0}{Ah}\left(\frac{\partial^2 E(B)}{\partial B^2}\right)_{B_0}, \quad (3.3)$$

h is the height of the BS, and it was defined as $h = \Delta z + 2R_{\text{vdW}}$; Δz is the puckering height of the sheet and R_{vdW} is the van der Waals radius.[15] A_0 and B_0 are the equilibrium lattice constants.

3.3.2 The Planar Boron Sheet

The optimized planar model (a) seems to form a triangular lattice with one atom per unit cell and a single lattice constant A, which is in the range of a typical boron-boron bond length $A = a_{\text{B-B}} = 1.69$ Å (see Fig. 3.10(b)). But within the accuracy of the given methods, we cannot really decide whether the lattice structure is perfectly triangular or slightly less symmetric. Assuming perfect triangular symmetry the two elastic moduli C_x and C_y are equal, and they are surprisingly big: $C_y = C_y \approx 750$ GPa, which is comparable to the ones in graphene (≈ 1 TPa). So even if the planar BS is metastable compared to other model boron sheets, it nevertheless has an extraordinary high stiffness. The electronic charge density is nearly uniform in the interstitial region, and the band structure (see Fig. 3.10(a)) is similar to the band structure of a free electron gas. These results seem to indicate some metallic bonding, as pointed out by Evans *et al.* [173], but such a picture cannot really account for the planarity and the high elastic modulus of this BS. A different qualitative picture of the chemical bonding is obtained after looking at the electron localization function [168] (ELF) in Fig. 3.10(b). Here we observe a simple network of two- and three-center bonds being less localized (ELF ≈ 0.7) than typical σ bonds (ELF ≈ 0.9), which are absent here. Thus the planar BS seems to be held together predominantly by multi–center bonds similar to the ones found in pure boron compounds. The chemical understanding of these bonds is still very limited. We think that, despite of its apparent metastability, model (a) could be an ideal theoretical tool to extend our present understanding of the nature of multi–center bonding in boron.

---

[15]The definition of h looks somewhat arbitrary, as a different definition for h or different van der Waals radii will certainly alter the values for the elastic moduli. But in a test calculation for a single graphite sheet, where we used $\Delta z = 0$ (no puckering) and $R_{\text{vdW}}^{\text{C}} = 1.7$ Å, we found $C_{11} = C_{22}$ to be 1.08 TPa, in excellent agreement with the literature values of $C_{11} = 1.06$ TPa [29]. For boron we would used $R_{\text{vdW}}^{\text{B}} = 1.7$ Å and $\Delta z = 0.82$ (see below and Table 3.3). Thus for model (a) we find that $h = 3.4$ Å, whereas for model (b) we find that $h = 4.22$ Å.

3.3. BROAD SHEETS

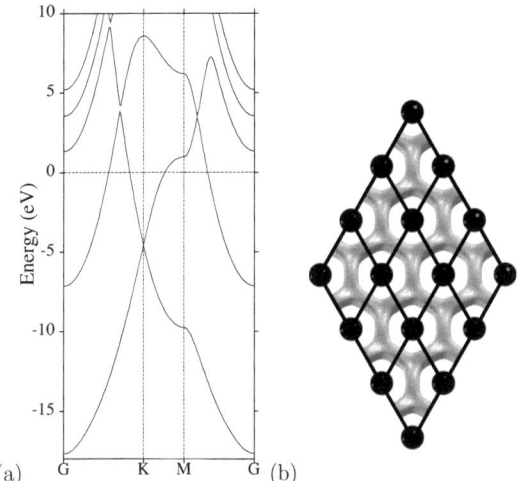

Figure 3.10: Properties of a planar boron sheet: (a) The two-dimensional band structure. (b) Black lines indicate the triangular unit cells, black spheres are boron atoms, and the orange (gray) contours show the electron localization function (ELF) at contours of 0.7. We observe a simple network of two- and three-center bonds.

3.3.3 The Puckered Boron Sheet

In Fig. 3.11 we show the unit cell of model (b). It consists of two basis atoms, and its planar projection is almost triangular. It is common to describe such a system with a face centered rectangular unit cell with lattice constants A and B. For $A/B = \sqrt{3} = 1.732$ a planar projection of the system would be triangular. In our case $A/B = 1.76$, which is a small, but noticeable departure from triangular symmetry. Due to a puckering of $\Delta z = 0.82$ Å, such a system can best be described using a three-dimensional orthorhombic unit cell. The corresponding lattice parameters and bond lengths can be found in Table 3.3.

In the following we will analyze the properties of model (b), which turns out to be the most stable structure for broad BS. Mind that from now on, whenever we write 'boron sheet' (BS), we will only refer to model (b).

In order to compare the BS with a known boron structure we also calculated the cohesive energy of α–rhombohedral boron, which turns out to be 7.51 eV/atom. This gives an energetic difference of 0.57 eV/atom (0.58 and 0.57 eV/atom in Refs. [173] and [174], respectively), which is huge, but one has to take into account that we are

Figure 3.11: The orthorhombic unit cell of model (b) with two basis atoms (see Table 3.3). In a xy-projection atom 1 is located at the corners of a rectangular unit cell, while atom 2 is located at the center of the unit cell. Along the z direction the boron atoms will generate a simple up and down puckering, with puckering heights around $\Delta z = 0.82$ Å.

comparing a single boron sheet with a bulk reference structure.

Mechanical Properties and Bonding

The elastic modulus of model (b) strongly depends on the stretching directions. In Table 3.3 we roughly find that $C_y \approx 2C_x$. How can one explain those rather obvious anisotropies?

To this end, let us have a look at the charge density of the BS (see Fig. 3.12). We clearly observe some parallel linear chains of σ bonds lying along the armchair direction. Their bond length is $a^\sigma_{B-B} = 1.60$ Å. At lower densities ($\rho < 0.7$ e/Å^3, not displayed) a largely homogeneous distribution with a rather complex shape appears, which may be assigned to multi–center bonding typical for boron materials. An analysis of the electron localization function [168] (ELF) leads to similar results, such that we obtain the following preliminary picture of the bonding: on a first level the sheet is held together by homogeneous multi–center bonds, but on a second level there are strong σ bonds lying only along the armchair direction.

Due to the strong σ bonds, any stretching of the BS along the armchair (= y) direction will be much harder than a similar stretching along its zigzag (= x) direction, where only the slightly weaker multi–center bonds are involved. These results are quite different from the results obtained by Evans et al., who conjecture that the σ bonds are strong but soft [173]. But here we clearly observe that the σ

3.3. BROAD SHEETS

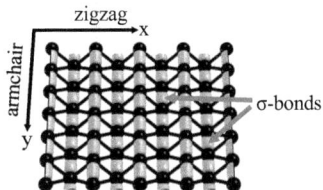

Figure 3.12: Orange (gray): charge density contours of the boron sheet (model (b)) at 0.9 e/Å^3. One observes parallel linear chains of sp hybridized σ bonds lying along the armchair direction.

bonds are strong and stiff. However other basic findings of Evans *et al.* are in good agreement with our results for planar and puckered BSs.

In general the elastic moduli involved are quite high; the stiffness of the σ bonds along the armchair direction is comparable to the stiffness of a graphene sheet. Furthermore the broken triangular symmetry of the BS's 2D lattice structure is another direct consequence of anisotropic bond properties.

Evans *et al.* also found that BNTs of different chiralities have different stiffnesses [173]. This can be confirmed by our bonding picture, although our results suggest that zigzag BNTs should be somewhat stiffer than armchair BNTs, while Evans *et al.* noted the opposite (the armchair and zigzag direction are swapped in their and our treatment, see Sec. 3.4.1). We thus conclude that the relation between the microscopic elastic modulus and the macroscopic Young's modulus must be rather complicated in the case of BS and BNTs.

Electronic Properties

The two-dimensional band structure of the BS $E^{\mathrm{BS}}(k_x, k_y)$ is plotted in Fig. 3.13 along lines of symmetry. The BS is metallic, as there are two bands crossing the Fermi energy, which is in perfect agreement with earlier studies of BSs [176, 177].

In order to find out about the hybridization of the σ bonds, we plotted the corresponding amount of s and p_y character indicated by the fatness of the bands.[16] We do not find individual dispersions of s or p bands, and the lowest lying bands show dispersions which *share* s and p_y character. That means they are bands consisting

---

[16]The orientation of the p_x, p_y, and p_z orbitals coincides with the orientation of the coordinate systems in Figs. 3.11 and 3.12.

CHAPTER 3. NOVEL PHASES OF ELEMENTAL BORON

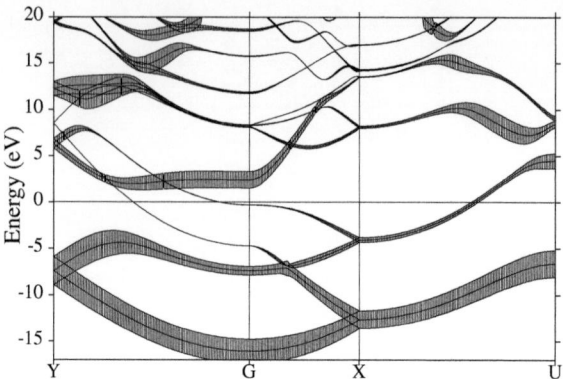

Figure 3.13: The band structure of the model BS. The fatness of the bands indicates their *sp* character, and it shows that the σ bonds in Fig. 3.12 must be of *sp* type. The Fermi energy E_F lies at $E = 0$, G is the Γ point. The position of the special points is given in Fig. 3.14.

of *sp* hybridized orbitals:

$$|sp_a\rangle = \frac{1}{\sqrt{2}}(|s\rangle + |p_y\rangle)$$
$$|sp_b\rangle = \frac{1}{\sqrt{2}}(|s\rangle - |p_y\rangle).$$

The directional coincidence of the p_y orbitals with the σ bonds in Fig. 3.12 identifies them to be of *sp* type. The strength of the σ bonds originates from the fact that the bands lie 5 to 15 eV below the Fermi energy.

The physical picture behind this multi–center bonding seems to be more complicated and it is still under investigation. So far we tried to analyze the multi–center bonds using a first nearest neighbor tight–binding model, which comprises the remaining p_x and p_z orbitals as basis states. But it turned out that this treatment can only partially reproduce the conduction bands in Fig. 3.13. Thus a larger basis set is needed.

In Sec. 3.3.1 we indicated that the puckering has a stabilizing effect for the BS. Now we are in a good position to explain this observation: any flattening of the BS would cause p_x orbitals to interfere with the σ bonds and eventually destroy them. And indeed, the analysis of the charge density and ELF of the planar BS in Sec. 3.3.2 showed that there are no σ bonds involved, but only multi–center bonds.

3.3. BROAD SHEETS

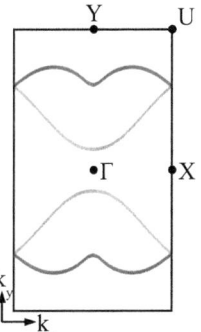

Figure 3.14: The two-dimensional Fermi surface of the boron sheet within the first Brillouin zone. It consists of two contours in red (black) and yellow (gray), which correspond to the two bands crossing the Fermi energy in Fig. 3.13.

Finally we want to show the two–dimensional Fermi surface $E_F = E^{\mathrm{BS}}(k_x, k_y)$ of the BS in Fig. 3.14. It obviously consists of two contours, which are dispersed throughout the Brillouin zone. This clearly shows the metallic properties of the BS.

3.3.4 Summary

In this section we studied a number of different structure models for broad boron sheets (BSs). All of them are metallic, and we found that for a 16 atom supercell, the model with a simple up-and-down puckering will be the most stable one. Large quasiplanar boron clusters with a similar structure (B_{22} [23], B_{48} [176], and B_{96} [147]) were already reported before. Now they may be understood as a first indication for the onset of periodicity in finite layered boron systems, and thus they are an independent confirmation of the current structure model.

A flat BS has a rather high stiffness, and the structure seems to be held together primarily by multi–center bonds (see Sec. 3.3.2). Although the sheet is less stable than previously known bulk phases of boron, as shown here and elsewhere [173, 174], the model sheet could be the ideal theoretical tool for studying complex multi–center bonds.

After describing the lattice structure of the stable BS, we have analyzed its band structure, the corresponding charge densities, and the electron localization function. This would lead to the following preliminary picture of the chemical bonding: on the one hand the sheet is held together by homogeneous multi–center bonds, on

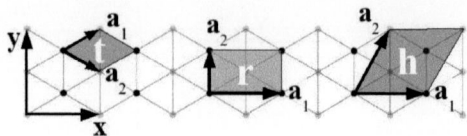

Figure 3.15: The triangular (t), the rectangular (r), and the honeycomb-derived (h) primitive cells that are used to characterize boron nanotubes. They contain one, two, and three atoms, respectively. Only the rectangular cell may properly describe the puckering of the boron sheet (indicated by black and gray atoms in the background).

the other hand there are linear sp hybridized σ bonds exclusively lying along the armchair direction of the sheet. The existence of sp hybridization in quasiplanar BS is somewhat surprising given the fact that earlier studies would always claim sp^2 hybridization. The rather anisotropic bond properties of the sheets lead to different elastic moduli C_x and C_y for stretching the BS in the x and in the y direction. Furthermore puckering of the BS, which breaks the triangular symmetry, may be understood as a key mechanism to stabilize the $sp\ \sigma$ bonds. Our results indicate that the sheet analyzed in this study is the boron analog of a single graphene sheet, a possible precursor of boron nanotubes (BNTs), and we wonder whether broad BSs might exist in nature.

3.4 Nanotubes

In the last section we determined a realistic model structure for a broad boron sheet being the boron analog of graphene. This model will be used now to predict the structure, the stability, the electronic and mechanical properties of BNTs. However, our first task will be to define a suitable classification scheme for BNTs based on the symmetry of the BS.

3.4.1 The Mathematical Description of Ideal Boron Nanotubes

Wrapping Vector

The geometrical construction of BNTs from a BS is similar to the construction of carbon nanotubes from a graphene sheet [29]: The basic tubular structure of BNTs is characterized by a wrapping vector W that defines one side of a rectangle. The second side T, being perpendicular to W, is chosen such that the rectangle can be periodically repeated along T (see Fig. 3.16). This rectangular patch is rolled up

3.4. NANOTUBES

to a cylinder such that $\boldsymbol{W}$ becomes the circumference of the nanotube with radius $R = |\boldsymbol{W}|/2\pi$. We will call any BNT, whose structure may be described by such a construction, an *ideal boron nanotube*.

Due to the fact that a proper structure model for BS was missing for a long time, there is some confusion in the literature about a proper reference lattice structure. In the work of Cabria *et al.* [174] and in earlier works by us [178, 177] (which is in full analogy to the construction of carbon nanotubes) the BNTs are related to a honeycomb lattice and the wrapping vector $\boldsymbol{W}^{\mathrm{h}}$ is defined as

$$\boldsymbol{W}^{\mathrm{h}} = (n, m) = n\boldsymbol{a}_1^{\mathrm{h}} + m\boldsymbol{a}_2^{\mathrm{h}}, \tag{3.4}$$

$\boldsymbol{a}_{1,2}^{\mathrm{h}}$ are the primitive vectors of a honeycomb lattice and n, m are integers. Here each unit cell has one additional atom at the center of the honeycombs, thus consisting of three rather than two atoms (see Fig. 3.15). Gindulyte *et al.* [25], Evans *et al.* [173], and Leys *et al.* [179] relate their BNTs to the simple triangular lattice, which only has one atom per unit cell:

$$\boldsymbol{W}^{\mathrm{t}} = (i, j) = i\boldsymbol{a}_1^{\mathrm{t}} + j\boldsymbol{a}_2^{\mathrm{t}}, \tag{3.5}$$

$\boldsymbol{a}_{1,2}^{\mathrm{t}}$ are the primitive vectors of a triangular lattice, and i, j are integers. $\boldsymbol{W}^{\mathrm{h}}$ and $\boldsymbol{W}^{\mathrm{t}}$ can be transformed into each other by using[17]

$$(n, m) \mapsto (i, j) = (n + 2m,\ n - m), \tag{3.6}$$

$$(i, j) \mapsto (n, m) = \frac{1}{3}(i + 2j,\ i - j). \tag{3.7}$$

From Fig. 3.15 we see that the two definitions are based on primitive vectors which have different orientations.[18] This leads to the rather unsatisfactory situation that armchair and zigzag directions are swapped in the two descriptions (see Table 3.4 for example). Cabria *et al.* found that all $(n, 0)$ zigzag and all $(2n, 2n)$ armchair BNTs have puckered surfaces, while the $(2n + 1, 2n + 1)$ armchair tubes shall be smooth due to the fact that an odd number of boron rows along the tube surfaces does not allow for the formation of the simple up and down puckering [174]. We think that these results are not an intrinsic property of BNTs, but rather a consequence of an unsuitable reference lattice system that is unable to properly describe the puckering of the boron sheet, see Fig. 3.15. Furthermore, the puckering breaks the hexagonal symmetry underlying the honeycomb and the triangular lattices.

---

[17] These relations are obtained from: $\boldsymbol{a}_1^{\mathrm{h}} = \boldsymbol{a}_1^{\mathrm{t}} + \boldsymbol{a}_2^{\mathrm{t}}$ and $\boldsymbol{a}_2^{\mathrm{h}} = 2\boldsymbol{a}_1^{\mathrm{t}} - \boldsymbol{a}_2^{\mathrm{t}}$.

[18] For the plot in Fig. 3.15 we used $A = \sqrt{3}B$. This induces triangular symmetry into a rectangular lattice, and the primitive vectors are $\boldsymbol{a}_1^{\mathrm{h}} = \boldsymbol{a}_1^{\mathrm{r}} = \sqrt{3}B(1, 0)$, $\boldsymbol{a}_2^{\mathrm{h}} = \sqrt{3}B(\frac{1}{2}, \frac{1}{2}\sqrt{3})$, and $\boldsymbol{a}_1^{\mathrm{t}} = B(\frac{\sqrt{3}}{2}, \frac{1}{2})$, $\boldsymbol{a}_2^{\mathrm{t}} = B(\frac{\sqrt{3}}{2}, -\frac{1}{2})$.

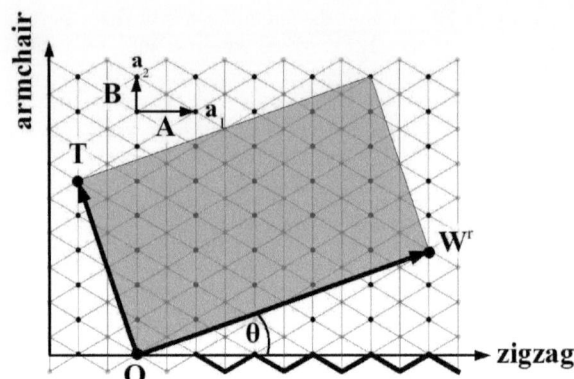

Figure 3.16: The geometrical construction of an *ideal* boron nanotube from a boron sheet: the red (gray) area is cut and rolled up such that W^r will become the circumference of the nanotube. O is the origin, W^r is the wrapping vector, T is the translational vector, θ is the chiral angle measured with respect to the zigzag direction, $a_{1,2}$ are the primitive vectors of the underlying *rectangular* lattice, and A and B are the lattice constants (see text). The puckering of the boron sheet is indicated by black and gray atoms shown in the background. The zigzag and the armchair directions are perpendicular to each other. This figure corresponds to $W^r = (5,3)$ and $A/B = \sqrt{3}$, which implies $T = (-1,5)$.

Therefore we think that all of these descriptions are inappropriate to classify BNTs. On the basis of the current BS model we would like to put forward a *different* way of describing BNTs, based on a rectangular lattice underlying the two-dimensional structure of the BS.

We define the wrapping vector W^r as

$$W^r = (k,l) = k a_1^r + l a_2^r, \qquad (3.8)$$

k,l are integers, and $a_1^r = A(1,0)$ and $a_2^r = B(0,1)$ are the primitive vectors of the rectangular lattice (see Figs. 3.15 and 3.11); A and B are the lattice constants from Table 3.3. In analogy to the Dresselhaus construction for carbon nanotubes [29], we define the chiral angle θ as the angle between the vectors W^r and a_1^r, i.e., θ is measured with respect to the zigzag direction that coincides with a_1^r (see Fig. 3.16).

The categorization of BNTs will be different from other classification schemes because the reduced symmetry of a BS increases the number of possible types of nanotubes. The range for the chiral angle is $0° \leq \theta \leq 90°$ and for the chiral indices

3.4. NANOTUBES

(k, l) we find that $k, l \geq 0$. Zigzag BNTs correspond to $\theta = 0°$ and $(k, l) = (k, 0)$, and armchair BNTs correspond to $\theta = 90°$ and $(k, l) = (0, l)$.[19]

$\boldsymbol{W}^{\mathrm{h}}$ and $\boldsymbol{W}^{\mathrm{t}}$ cannot directly be converted to $\boldsymbol{W}^{\mathrm{r}}$, as they are defined for lattices with different symmetries. For the achiral types, one can use the following analogies (examples are listed Table 3.4)

$$\begin{aligned}
\text{zigzag:} \quad (k, 0)^r &\leftrightarrow (k, 0)^h \\
&\leftrightarrow (k, k)^t, \\
\text{armchair:} \quad (0, l)^r &\leftrightarrow (l/3, l/3)^h \\
&\leftrightarrow (l, 0)^t.
\end{aligned} \tag{3.9}$$

Translational Vector

The *tubular* unit cell of an ideal BNT, being the red (gray) area in Fig. 3.16, may be defined properly by a wrapping vector $\boldsymbol{W}^{\mathrm{r}}$ (Eq. (3.8)) and the so-called translational vector $\boldsymbol{T}$, which is perpendicular to $\boldsymbol{W}^{\mathrm{r}}$:

$$\begin{aligned}
\boldsymbol{T} = (t_1, t_2) &= t_1 \boldsymbol{a}_1^{\mathrm{r}} + t_2 \boldsymbol{a}_2^{\mathrm{r}}, \\
t_1 &= \begin{cases} -\text{numerator}(f) & : k \neq 0 \\ 1 & : k = 0 \end{cases} \\
t_2 &= \begin{cases} \text{denominator}(f) & : k \neq 0 \\ 0 & : k = 0 \end{cases} \\
f &= \text{reduce}\left(\frac{lB^2}{kA^2}\right).
\end{aligned} \tag{3.10}$$

t_1, t_2 are integers and reduce (r) should indicate that the fraction r must be reduced before determining its numerator and denominator.

Let us consider the length of the translational vector $\boldsymbol{T}$. For the achiral BNTs $|\boldsymbol{T}|$ is particularly small: for all $(k, 0)$ zigzag types we have $\boldsymbol{T} = (0, 1)$, and for $(0, l)$ armchair BNTs $\boldsymbol{T} = (1, 0)$. For the chiral types $\boldsymbol{T}$ depends on the ratio B^2/A^2 (see the last line of Eq. (3.10)). Using $A = 2.819$ and $B = 1.602$ we obtain reduce $(B^2/A^2) = 2566404/7946761$. Therefore the coefficients t_1 and t_2 are really huge numbers, which means that $|\boldsymbol{T}|$ becomes macroscopically large. For A and B chosen as above, the estimated length of $\boldsymbol{T}$ for all chiral BNTs will be in the mm range. Imposing some additional symmetry constraints by relating the lattice constants will immediately remedy this problem. For example after choosing $A =$

---

[19]Restricting the chiral angle θ to that range implies that for chiral nanotubes the indices (k, l) actually specify an enantiomeric pair (optical isomers), i.e., nanotubes that are mirror images of each other. These enantiomers have opposite chiralities (right-handed versus left-handed) and different line group symmetries [180, 181]. This non–unique classification is in analogy to the classification of CNTs.

$\sqrt{3}B$, fraction$(B^2/A^2) = 1/3$, i.e., $|\boldsymbol{T}|$ will be reduced to just a few lattice constants (this case was used to generate Figs. 3.16 and 3.17). Quite obviously the chiral BNTs the specific ratio B^2/A^2 determines the length of the translational vector.

Note that boron compounds usually have a whole set of different B–B bond lengths, which means that boron does not necessarily favor highly symmetric arrangements. The bond lengths are more flexible than for typical covalent elements like carbon, and the lattice constants A and B of the BS cannot be seen as fixed parameters; they will have slightly different values in BNTs. Furthermore, the broken planar triangular symmetry of the BS is rather typical for boron, and we should expect that for ideal chiral BNTs, even with different values of A and B, the translational vector might still be rather huge.

To summarize: any departure from the triangular symmetry in the BS will create chiral BNTs, which have macroscopically large translational vectors, and achiral types, where $|\boldsymbol{T}|$ is of the order of the lattice constants. Thus achiral BNTs (armchair and zigzag) have a one-dimensional translational symmetry along the tube's axis, which is not present in chiral BNTs. For the latter it might be better to think in terms of helical (chiral) symmetries only. Therefore we predict the existence of *helical currents* in ideal chiral BNTs. Such currents could lead to very interesting physical effects such as strong magnetic fields [182] and self-inductance effects leading to an inductive reactance [183] of chiral BNTs.

Band Structure

Within the limit of big nanotube radii, where curvature effects are small, one may derive the one-dimensional band structure of an ideal BNT $E_\mu(k')$ by a zone-folding technique [29], starting from the two-dimensional band structure of a BS $E^{\text{BS}}(k_x, k_y)$. The zone-folding is usually based on the translation symmetry of nanotubes along their axis. However, given the absence of translational symmetry in ideal chiral BNTs, we have to base our zone-folding theory on the helical symmetry of BNTs [184, 185].

Figure 3.17 illustrates that besides constructing a BNT by repeating a *tubular* unit cell one can also build a nanotube by repeating a *helical* unit cell along a spiral winding around the surface of the tube. The direction of this spiral is given by the helical vector $\boldsymbol{H}$ [184, 185] (in Ref. [29] it is called the symmetry vector $\boldsymbol{R}$). Uncoiled into a plane, this vector defines the direction of a translational symmetry (see Eq. (3.11) and thereafter). The helical unit cell is specified by $\boldsymbol{H}$ and the vector $\boldsymbol{K} \perp \boldsymbol{H}$. Furthermore, we define a vector $\boldsymbol{L} \parallel \boldsymbol{H}$, such that $\boldsymbol{W}^{\tau} = (k, l) = \boldsymbol{K} + \boldsymbol{L}$ (see Fig. 3.17).

3.4. NANOTUBES

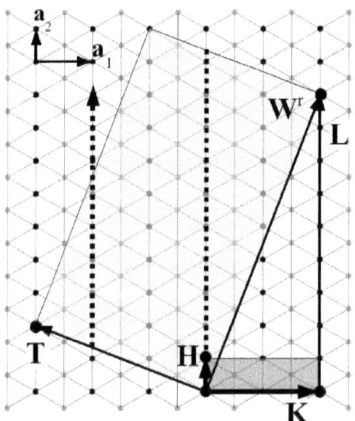

Figure 3.17: Two different ways of "building up" a nanotube: the *tubular* unit cell in light gray (see also Fig. 3.16) is repeated along the nanotube's axis, which lies parallel to $\boldsymbol{T}$. The *helical* unit cell in red (dark gray) is translated along spirals (represented by the dotted lines) on the surface of the nanotube; it is defined by the helical vector $\boldsymbol{H}$ and vector $\boldsymbol{K}$. It holds that $\boldsymbol{W}^{\mathrm{r}} = \boldsymbol{K} + \boldsymbol{L}$. Here $\boldsymbol{H} = (0, 1)$, $\boldsymbol{K} = (2, 0)$ and $\boldsymbol{L} = (0, 9)$, and therefore $\boldsymbol{W}^{\mathrm{r}} = (2, 9)$. The length of $\boldsymbol{T} = (-3, 2)$ was artificially reduced by choosing $A/B = \sqrt{3}$.

The helical wave functions are restricted by the following criteria:

$$\Psi_{\mu k'}(\boldsymbol{r} + \boldsymbol{H}) = \Psi_{\mu k'}(\boldsymbol{r}) \exp(ik'|\boldsymbol{H}|), \qquad (3.11)$$

$$\Psi_{\mu k'}(\boldsymbol{r} + \boldsymbol{W}^{\mathrm{r}}) = \Psi_{\mu k'}(\boldsymbol{r}). \qquad (3.12)$$

Equation (3.11) defines a one-dimensional Bloch state with $-\pi/|\boldsymbol{H}| < k' < \pi/|\boldsymbol{H}|$ and imposes the condition that k' has to be parallel to the reciprocal lattice vector related to $\boldsymbol{H}$. Equation (3.12) is the tubular boundary condition. In order to construct the helical wave functions $\Psi_{\mu k'}$ we use the wave functions of the BS $\Psi_{\boldsymbol{k}}^{\mathrm{BS}}(\boldsymbol{r})$ which have the Bloch property:

$$\Psi_{\boldsymbol{k}}^{\mathrm{BS}}(\boldsymbol{r} + \boldsymbol{R}) = \exp(i\boldsymbol{k} \cdot \boldsymbol{R}) \Psi_{\boldsymbol{k}}^{\mathrm{BS}}(\boldsymbol{r}), \qquad (3.13)$$

where $\boldsymbol{R}$ is a vector of the Bravais lattice formed by $\boldsymbol{a}_1^{\mathrm{r}}$ and $\boldsymbol{a}_2^{\mathrm{r}}$. Since the vectors $\boldsymbol{H}$ and $\boldsymbol{W}^{\mathrm{r}}$ are elements of the same Bravais lattice, Eq. (3.11) will automatically be satisfied, and Eq. (3.12) together with Eq. (3.13) will yield

$$1 = \exp\left[i(\boldsymbol{k} \cdot \boldsymbol{W}^{\mathrm{r}})\right]. \qquad (3.14)$$

In order to proceed, we now choose a direction for $\mathbf{H}$, which may be any Bravais lattice vector.[20] By choosing $\mathbf{H} = \mathbf{T}$ we recover the case of a tubular unit cell, as described above and in Ref. [29]. But in order to make the calculation as simple as possible we assign $\mathbf{H} = \mathbf{a}_2^r = (0,1)$. Then it follows that $\mathbf{K} = (k,0)$ and $\mathbf{L} = (0,l)$ (see Fig. 3.17). As $\mathbf{H} \parallel y$ we have to choose $k' = k_y$. After inserting Eq. (3.14) into $E^{\mathrm{BS}}(k_x, k_y)$ we finally obtain the zone-folded band structure of ideal (k,l) ($k \neq 0$) BNTs as

$$E_\mu^{(k,l)}(k') = E^{\mathrm{BS}}\left(\frac{2\pi}{kA}\mu - \frac{lB}{kA}k', k'\right), \tag{3.15}$$
$$-\frac{\pi}{B} < k' < \frac{\pi}{B},$$
$$\mu = 0, \cdots, k-1$$

Equation (3.15) will break down for $(0,l)$ armchair BNTs, due to a chiral index $k = 0$. But as mentioned before, we are free to change the direction of $\mathbf{H}$, and in such a case we use $\mathbf{H} = \mathbf{a}_1^r = (1,0)$ and have $k' = k_x$. We then obtain

$$E_\mu^{(0,l)}(k') = E^{\mathrm{BS}}\left(k', \frac{\mu}{l}\frac{2\pi}{B}\right), \tag{3.16}$$
$$-\frac{\pi}{A} < k' < \frac{\pi}{A},$$
$$\mu = 0, \cdots, l-1$$

To decide whether a certain ideal BNT is metallic or not we can zone–fold the BS's Fermi surface given in Fig. 3.14. We did so and found that *all* ideal BNTs are indeed metallic, irrespective of their radius and chiral angle. The only ideal BNTs that are not metallic are the (0,1) and (0,2) types. But these structures have a very small radius and are thus highly unrealistic and we can safely rule them out, as they are not even covered by the Aufbau principle [22].

3.4.2 Real Boron Nanotubes

After defining a classification scheme for BNTs and analyzing the basic properties of ideal BNTs in the preceeding section, we will now simulate BNTs and try to figure out whether there is a relation between the *ideal* BNTs, derived theoretically from the model BS, and *real* BNTs. Similar to the BSs, the basic structure of BNTs is given by the Aufbau Principle (see Sec. 3.2.3) but the surface puckering remains unspecified and has to be determined using *ab initio* simulations.

---

[20]In the most general case the zone-folded band structure of ideal BNTs is given by $E_\mu(k') = E^{\mathrm{BS}}(k'\mathbf{G_H}/|\mathbf{G_H}| + \mu\mathbf{G_K})$, with $\mathbf{G_H}$ and $\mathbf{G_K}$ being the reciprocal lattice vectors of $\mathbf{H}$ and $\mathbf{K}$, respectively, $-\pi/|\mathbf{H}| < k' < \pi/|\mathbf{H}|$, $\mu = 0, \ldots, N-1$, and $N = |\mathbf{H} \times \mathbf{K}|/|\mathbf{a}_1^r \times \mathbf{a}_2^r|$ [29].

3.4. NANOTUBES

(k,l)	$(n,m)/(i,j)$	n	Isom.	C_j	$a^{\text{axial}}_{\text{B-B}}$	$a^{\text{diagonal}}_{\text{B-B}}$	$a^{\text{circumferential}}_{\text{B-B}}$	$\bar{R} \pm \Delta R$	$E^{\text{ind}}_{\text{coh}}$	$E^{\text{rope}}_{\text{coh}} - E^{\text{ind}}_{\text{coh}}$
(9,0)	(9,0)/(9,9)	18	α	C_3	1.61^σ	1.77,1.83,1.86		3.86 ± 1.09	6.93	+0.07
			β	C_1	1.61^σ	1.67 – 1.87			6.92	
			γ	C_3	1.61^σ	1.81,1.82		3.83 ± 0.51	6.91	+0.04
			δ	C_9	1.61^σ	1.83		4.17 ± 0.39	6.83	
			ε	C_3	1.64^σ	1.67,1.81		4.39 ± 0.29	6.78	
(10,0)	(10,0)/(10,10)	20	α	C_2	1.60^σ	1.79,1.81,1.82,1.87		3.84 ± 1.97	6.91	+0.01
			β	C_2	1.61^σ	1.82,1.83,1.84		4.08 ± 1.18	6.90	+0.07
			γ	C_{10}	1.61^σ	1.83		4.60 ± 0.41	6.85	
(12,0)	(12,0)/(12,12)	24	α	C_6	1.61^σ	1.73,1.83,1.85		5.05 ± 0.65	6.90	+0.02
			β	C_{12}	1.61^σ	1.82		5.48 ± 0.41	6.87	+0.05
(0,12)	(4,4)/(12,0)	24	α	C_6		1.69	$1.59^\sigma,1.69,1.85$	2.64 ± 0.68	6.68	+0.3
(0,18)	(6,6)/(18,0)	36	α	C_6		1.70,1.74	$1.56^\sigma,1.60^\sigma,1.71,1.75$	4.48 ± 0.57	6.74	+0.27
			β	C_{18}		1.75	$1.53^\sigma,1.76$	4.74 ± 0.34	6.72	
(0,24)	(8,8)/(24,0)	48	α	C_6		1.74,1.75	$1.54^{\sigma,i},1.57^{\sigma,i},1.64^{\sigma,o},1.72,1.74$	5.99 ± 0.58	6.81	+0.3

Table 3.4: Structural data and energies of different isomers of free-standing boron nanotubes: (k,l), (n,m), (i,j): different chiral indices for the same tube type (see Sec. 3.4.1); n: number of atoms per unit cell; Isom.: label of isomer; C_j: rotational symmetry; $a^{\text{axial}}_{\text{B-B}}$, $a^{\text{diagonal}}_{\text{B-B}}$, $a^{\text{circumferential}}_{\text{B-B}}$: boron-boron bond lengths in axial, diagonal and circumferential direction of a nanotube, the superscript σ indicates that this bond is a σ bond, superscripts σ, i and σ, o refer to inner and outer rings, respectively; $\bar{R} \pm \Delta R$: mean radius of a nanotube (Eq. (3.17)) and maximal radial variation (Eq. (3.18)); $E^{\text{ind}}_{\text{coh}}$: cohesive energy of a free-standing (individual) nanotube (Eq. (3.1)); $E^{\text{rope}}_{\text{coh}} - E^{\text{ind}}_{\text{coh}}$: this energy is gained when the same nanotube is arranged in a bundle (rope). All energies are given in eV/atom and all lengths are given in Å.

Therefor we started from a series of initial structures with smooth surfaces, which were optimized in a triangular BNT bundle (rope). Here the strong tube–tube interactions (see Sec. 3.4.2) distort the surfaces and naturally induce some puckering. The energy of this configuration is $E_{\text{coh}}^{\text{rope}}$. In order to simulate free–standing (individual) BNTs we then increased the intertubular distance to 6.4 Å, and optimized those configurations while keeping the intertubular distances fixed. The energy here is $E_{\text{coh}}^{\text{ind}}$ ($E_{\text{coh}}^{\text{rope}}$ and $E_{\text{coh}}^{\text{ind}}$ are defined after Eq. (3.1)). Note that such an approach does not impose any surface puckering which is solely found in the numerical simulations.

All free–standing BNTs are shown in Figs. 3.18, 3.19, and 3.22 and the structural data and energies are collected in Table 3.4. Apart from their bond lengths and rotational symmetries we also listed the geometrical mean radius of each tube $\bar{R}$, as well as the maximal radial variation ΔR, defined as:

$$\bar{R} = \frac{R^{\min} + R^{\max}}{2}, \qquad (3.17)$$

$$\Delta R = R^{\max} - \bar{R} = \bar{R} - R^{\min}, \qquad (3.18)$$

where $R^{\min}$ and $R^{\max}$ are the distances of the innermost and the outermost atoms from the center of the nanotube, respectively.

For many (k,l) BNTs we found more than just one isomer. Therefore each BNT was also given a Greek index which labels different isomers. The latter were ordered according to their cohesive energies, i.e., $(k,l)\alpha$ will denote the most stable isomer, $(k,l)\beta$ would be less stable, and so on.

Free Standing Nanotubes vs Nanotube Ropes

In Table 3.4 the "inter-tubular energy" $E_{\text{coh}}^{\text{rope}} - E_{\text{coh}}^{\text{ind}}$ is the energetic difference between a free–standing BNT and its bundled counterpart. One can see that it varies significantly from tube to tube. The intertubular energy seems to depend quite strongly on the structure type, the relative orientations of adjacent tubes in a rope, and on the specific type of surface puckering. Furthermore, the intertubular distance in different bundles, which was defined as the minimal separation between two apex atoms on adjacent nanotubes, varies between 1.7 and 3.5 Å.

It is obvious that the tube–tube interaction in BNT bundles (ropes) is completely different from what is known from carbon nanotubes, where the intertubular interaction is of van der Waals type. The latter is certainly much weaker, independent of the various structure types, and the intertubular distances are always around 3.4 Å. BNTs on the other hand *may* have *covalent* intertubular bonds [178, 186], and this leads to a sizeable intertubular bonding energy that depends quite strongly on structural details.

It is interesting to note that the intertubular energy of $(0,l)$ BNTs (armchair types) is significantly higher than for $(k,0)$ BNTs (zigzag). Below we will try to give

3.4. NANOTUBES

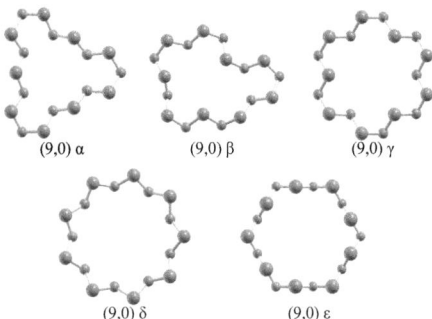

Figure 3.18: The cross sections of different isomers of a free–standing (9,0) zigzag boron nanotube. The big spheres represent the upper atoms and the small ones the lower atoms (with respect to the direction of the tube axis). The α and γ isomers are the free–standing counterparts of the (9,0)C and (9,0)B tubes in Ref. [178], respectively.

an explanation for this rather complex bonding scenario.

At this point, it is worth noting that the original motivation for this work was a recent study by ourselves, where we reported bundled zigzag BNTs that were somewhat *constricted* [178] (we define the concept of *constriction* at the end of Sec. 3.4.2). We conjectured that this constriction would most likely be caused by the arrangement of the tubes in a bundle, where the tube-tube interactions will force the tubes to have geometrical shapes different from free–standing BNTs. The free–standing counterparts of the constricted (9,0)C and (10,0)C BNTs from Ref. [178] are the (9,0)α[21] isomer in Fig. 3.18 and (10,0)β in Fig. 3.19. To our surprise the constriction did *not* disappear after isolating the tube. And even after substantially deforming the (9,0)α structure by homogeneous shrinking, by blowing it up, or by randomly elongating atoms out of their equilibrium position with a maximum amplitude of 0.3 Å, the free–standing (9,0)α BNTs always relaxed to their constricted forms. This finding is in clear contrast to our previous hypothesis, and it raises the important question where those constrictions finally come from. We will try to give an answer to this question in Sec. 3.4.3.

The Structure of Free Standing Boron Nanotubes

---
[21] For the (9,0)α BNT the intertubular distance was increased to only 4 Å.

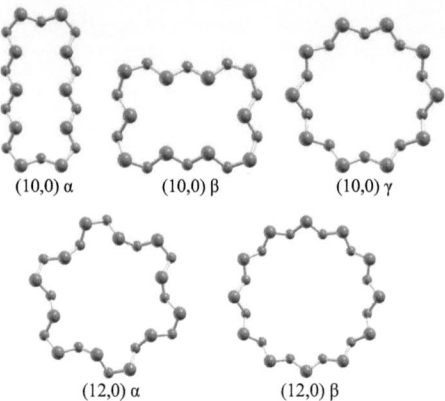

Figure 3.19: Cross–sectional view of various isomers of free–standing (10,0) and (12,0) zigzag boron nanotubes. Again the big spheres mark the upper atoms and the small ones mark the lower atoms. The (10,0)α and (10,0)β isomers are free–standing counterparts of the (10,0)B and (10,0)C structures in Ref. [178], respectively.

Zigzag Nanotubes For zigzag BNTs we found various isomers. Any zigzag BNT may be seen as a BS that was rolled up along its zigzag direction (see Fig. 3.12 or 3.16). Thus the linear chains of σ bonds will lie along its axial direction and they will remain straight. These basic bonding properties were typical for *all* zigzag BNT that we studied so far. We just show two typical examples in Fig. 3.20. Here the bond length of the σ bonds is quite similar to the bond length in the BS, as $a^{\sigma}_{B-B} = a^{axial}_{B-B} = 1.61$ Å (the only exception we found was (9,0)ϵ).

The tubes (9,0)δ, (10,0)γ, and (12,0)β are *ideal* BNTs, which denotes the fact that they were initially constructed by a "cut and paste" procedure described in Sec. 3.4.1, and then re–optimized using *ab initio* methods. Their structure is highly symmetric and we find two bond lengths, which are almost identical to the bond length in the BS. The puckering height $\Delta z = 2\Delta R \approx 0.8$ is also quite similar to the BS.

However, an ideal BNT does not seem to be the ground state of a real zigzag BNT, and we found less symmetric isomers that were higher in cohesive energy. It should be noted that zigzag tubes with a smooth surface were not considered here because their cohesive energies are significantly lower than those of puckered BNTs. As an example for the complex shape of zigzag BNTs one may study (9,0)ϵ, which

3.4. NANOTUBES

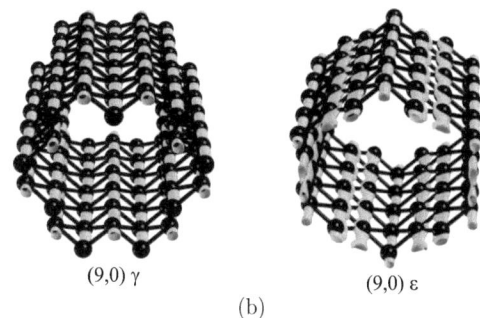

(a) (9,0) γ (b) (9,0) ε

Figure 3.20: Zigzag boron nanotubes and the presence of straight σ bonds along their axial direction, which are indicated by orange (gray) charge density contours at 0.9 e/Å^3. Due to a lack of stiff σ bonds along the circumferential direction, this type of nanotube might not be stable.

is the least stable isomer of all zigzag BNTs. (9,0)ε has a hexagonal cross section, which probably caused by its placement into a triangular supercell. Its facets may be seen as parts of a flat BS, whereas the corner pieces are parts of a puckered BS. From Fig. 3.20 we notice that the σ bonds along the sides are slightly more delocalized than the ones located at the corners. This means that any flattening would destabilize the sigma bonds, and the whole tube is highly metastable. (A similar but square-like structure was found by Evans et al. [173], which they labeled $(i, j) = (6, 6)$, but we think that this structure is highly metastable as well.) Thus the question will no longer be if zigzag BNTs are puckered, but *how* they are puckered.

The cross sections of the isomers with a high cohesive energy may be built from

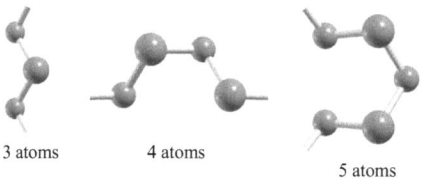

3 atoms   4 atoms   5 atoms

Figure 3.21: Three basic structure elements. The cross sections of the most stable zigzag boron nanotubes in Figs. 3.18 and 3.19 may be composed of these elements only.

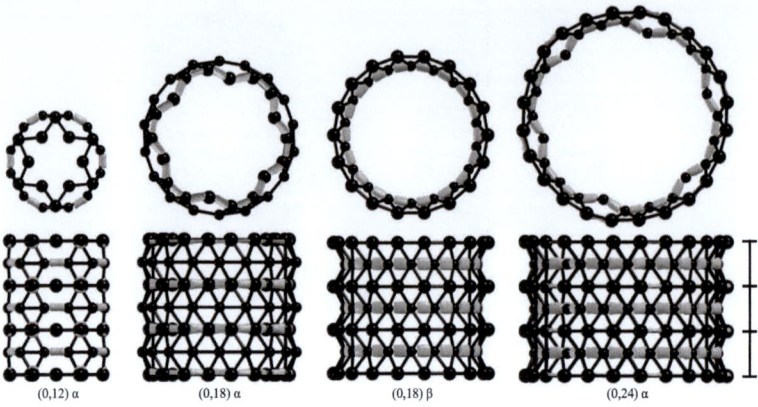

Figure 3.22: Top and side view of various free standing armchair boron nanotubes and the presence of σ bonds, which are indicated by orange (gray) charge density contours at 0.95 e/Å^3. All armchair nanotubes have bent σ bonds along the circumferential direction, which basically generate the strain energies of the tubes. The black bar on the right indicates the height of a supercell in axial direction that was used for our simulations; for esthetical reasons we actually displayed three identical units cells.

three basic structure elements that are shown in Fig. 3.21. The three-atomic structure element is directly related to the puckering of the BS (compare Fig. 3.12) whereas the four- and five-atomic elements are just special combinations of three-atomic structure elements. We see that the structure of zigzag BNTs is strongly related to the *local* structure of the puckered BS, but their *general* cross-sectional geometries seem to be more complicated and less symmetric than in the case of an ideal BNT. This loss of symmetry can also be extracted from the spectrum of diagonal bond lengths, which are associated with multi-center bonds. Rather than being equal to $a_{\text{B-B}}^{\text{diagonal}}$ of the BS (see Table 3.3) those bond lengths span a whole range $a_{\text{B-B}}^{\text{diagonal}} \approx 1.7 - 1.9$ Å.

Some of the most interesting structures are $(9,0)\alpha$, $(9,0)\beta$, and $(10,0)\alpha$, which have cross sections that are far from being circular. Nonetheless they exhibit high cohesive energies. Because of the observed unusual shapes of zigzag BNTs we assume that the multi-center bonds obviously possess a high directional flexibility, but at the same time they are also very stiff ($C_x = 0.42$ TPa in Table 3.3). Therefore it seems as if these bonds have some joint-like properties, i.e., they are easy to turn, but hard

3.4. NANOTUBES

to tear.

In the following we will call a zigzag BNT *constricted*, if it is composed of several five-atomic structure elements like the $(9,0)\alpha$ and the $(10,0)\beta$ isomers in our work. A constricted zigzag BNT was also found by Evans et al. [173], where it is labeled as a $(i,j) = (8,8)$ nanotube, and it corresponds to our $(10,0)\beta$ structure without the two horizontal three-atomic elements.

Armchair Nanotubes When rolling up a BS along its armchair direction, the puckered sheet (see Fig. 3.12) will be transformed into a tube that has inner and outer rings, and the σ bonds will lie along its circumferential direction. On the outer rings the length of the σ bonds will be increased and on the inner rings their length will be reduced. In Fig. 3.22 we see that for three systems discussed in this study, the σ bonds do really lie along the circumferential direction, and for the $(0,18)$ and $(0,24)$ systems an inner and an outer ring can clearly be identified.

In contrast to zigzag BNTs we did *not* find several isomers for the armchair types. Furthermore, we just discuss one ideal BNT, which is the $(0,18)\beta$ isomer. In analogy to zigzag BNTs, we found that this ideal BNT corresponds to a local energetic minimum, and the $(0,18)\alpha$ isomer of lower symmetry is 0.02 eV/atom more stable.[22] The latter has σ bonds solely along the inner rings, where the bond lengths are 1.56 and 1.60 Å. Along the outer ring, where the B-B distances (1.71 and 1.75 Å) are significantly longer, the curvature effect has destroyed the σ bonds.[23]

The $(0,24)$ system has similar properties, but here the curvature is smaller, and there are six additional weak σ bonds along the outer rings with a bond length of 1.64 Å. For even larger radii we expect the outer rings of armchair BNTs to develop σ bonds between every single atom.

The radius of the $(0,12)$ BNT is quite small, which makes it extremely difficult for the structure to align its σ bonds. We see that this tube possesses a different geometry, and even along the stiffer rings there are six instead of 12 σ bonds. It is obvious that for armchair BNTs with smaller and smaller radii, the curvature effect will successively destroy the circumferential σ bonds. For the smallest possible BNTs we should find no σ bonds at all, and the surface of the tube will become smooth. This agrees with earlier studies by ourselves [24, 176] and with the work of Evans et al. [173], where some armchair BNTs of small radii were studied and found to be smooth.

Any destruction of circumferential σ bonds within armchair BNTs of small radii

---

[22]The C_6 symmetry of all α isomers is probably not an intrinsic property, but rather caused by the fact that they were simulated in a triangular supercell.

[23]The diagonal bond lengths $a_{B-B}^{diagonal}$ (which connect the inner and the outer rings) are always shorter compared to the BS (compare Tables 3.3 and 3.4); we found them to be in the range $a_{B-B}^{diagonal} \approx 1.69 - 1.75$ Å.

will release electrons that can alter their chemical properties. In Sec. 3.4.2 we observed that the intertubular energy for armchair BNT ropes is much higher than for zigzag BNT ropes. Now a possible explanation would be that the released electrons in armchair BNTs induce an enhanced reactivity. In a rope of BNTs, this enhanced reactivity will lead to strong intertubular bonding for armchair BNTs of small radii. In zigzag BNTs the reactivity is lower, as a maximum number of σ bonds can always be achieved, due to the fact that curvature effects will not be able to weaken the axial σ bonds. Therefore we hypothesize that small sized armchair BNTs will have a higher reactivity than zigzag BNTs, and that this reactivity will further decrease with increasing radii.

This reactivity, which leads to the formation of intertubular bonds in BNT ropes, could be very useful when trying to embed BNTs into polymers [173], where strong chemical bonds between the nanotubes and the polymer matrix are needed in order to improve the mechanical properties of the composite.

To summarize, we have seen that *real* BNTs have striking similarities to *ideal* BNTs, which are derived from the puckered BS found in Sec. 3.3. This firmly establishes that the puckered BS is the precursor of BNTs. However, ideal and real BNTs differ in two points: real BNTs are less symmetric than ideal ones and the zigzag types can be constricted.

3.4.3 Strain Energy

Let us now compare the cohesive energy of every BNT ($E_{\text{coh}}^{\text{ind}}$ from Table 3.4) with the cohesive energy of the puckered BS ($E_{\text{coh}}^{\text{BS}}$ from Table 3.3). This energy difference will be called strain energy:

$$E_{\text{strain}}(k,l) = E_{\text{coh}}^{\text{BS}} - E_{\text{coh}}^{\text{ind}}(k,l). \tag{3.19}$$

It is the amount of energy that is needed to roll up a BS into a BNT. The microscopic origin of the strain energy in nanotubes are bent σ bonds along the circumferential direction of the tubes. These bonds have a strong tendency to jump back into a straight orientation, which generates a tension that may thus be quantified by the strain energy of the systems. Such a tension will stabilize the tubular shape, or to put it more clearly: it will make the nanotube round.

The strain energies of different (k,l) BNTs as a function of their mean radii (Eq. (3.17)) are plotted in Fig. 3.23. For the sake of comparison we also show the universal strain energy curve for carbon nanotubes. We call it universal because the strain energy only depends on the radius, but not on the chiral angle (chirality) of the nanotubes: $E_{\text{strain}}^{\text{C}} = E_{\text{strain}}^{\text{C}}(R)$.

As the BNTs are all puckered, there is some variability in the proper choice of a mean tubular radius. Since the strain energy is related to the position of the σ

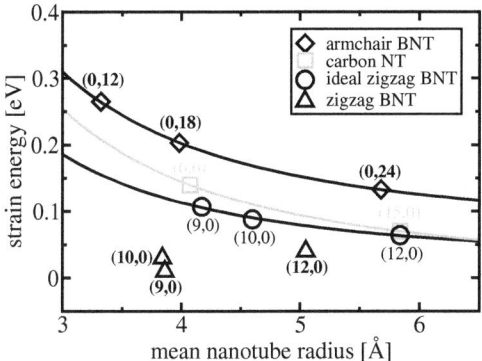

Figure 3.23: The strain energy of α isomers as a function of the mean radius $\bar{R}$ (Eq. (3.17)); for armchair boron nanotubes we used $\bar{R}_\sigma$ (Eq. (3.20)). In orange (gray) we show the universal strain energy curve for carbon nanotubes (□); the energy obviously depends on their radii, but not on their chiral angles. For *armchair* boron nanotubes (◇) we find a similar curve, but those boron tubes have more strain energy. For *zigzag* boron nanotubes (△) we cannot really plot a strain energy curve, as different nanotubes of different radii are almost isoenergetic. *Ideal* zigzag boron nanotubes (○) have less strain energy than their armchair counterparts, but they are metastable.

bonds, it makes sense to define the mean radius of armchair BNTs as

$$\bar{R}_\sigma = \frac{R_\sigma^{\min} + R_\sigma^{\max}}{2}. \qquad (3.20)$$

Here $R_\sigma^{\min}$ and $R_\sigma^{\max}$ are the distances of the innermost and the outermost atoms sharing σ bonds, which is measured from the center of the nanotube.

Boustani et al. [177] studied the elasticity of *armchair* BNTs with a tight-binding method and reported a typical strain energy curve lying below the one of carbon nanotubes.[24] Now, using an *ab initio* method, we also found that armchair BNTs have strain energy, but the latter is higher than for carbon nanotubes.

Different *ideal* zigzag BNTs in Fig. 3.23 have rather low strain energies. Here none of the σ bonds has to be bent, and the strain energy should be caused by the multi-center bonds. But those ideal BNTs are metastable, and isomers of lower symmetry have higher cohesive energies. Thus for the zigzag α isomers no strain energy curve may be plotted. They are more or less isoenergetic. It seems that zigzag

---

[24] In Ref. [177] smooth BNTs and a flat BS were compared.

BNTs can release some or all of their strain energy by lowering their symmetry and undergo internal deformations (see also Ref. [173]), possibly mediated by the joint-like properties of the multi-center bonds.

In summary we see that the strain energy in BNTs is mainly caused by bent σ bonds lying entirely (armchair) or only partially (chiral BNTs) along the circumferential direction. The multi-center bonds are always present, but they seem to have no serious influence. The apparent *absence* of strain energy in zigzag BNTs is caused by the fact that the linear σ bonds lie along the axial direction, only. But without smoothing bonding strains, the zigzag tubes are free to take a multitude of cross-sectional morphologies. This explains the number of different isomers that we found for (9,0), (10,0), and (12,0) zigzag BNTs and their bizarre shapes. The constriction of zigzag BNT, first reported in Ref. [178], is a clear consequence of the absence of strained bonds within zigzag BNTs. Armchair BNTs, which are geometrically stabilized by their strain energy, do not seem to have this kind of isomerism.

Chiral BNTs may be pictured as a certain combination of structural elements from armchair and zigzag tubes characterized by a chiral angle. Therefore we suppose that there will be a separate strain energy curve for every chiral angle lying between the armchair and the zigzag curves. The strain energies themselves will depend on the radii and on the chiral angle of a BNT: $E^B_{strain} = E^B_{strain}(R, \theta)$. This seems to be a unique property among all nanotubular materials reported so far. A profound analysis of strain energies in nanotubes will be given in Chapter 4.

But it remains open whether the strain energy of zigzag BNT will be completely absent, or just significantly smaller than for armchair BNTs. The present results are in favor of the former hypothesis. As carbon nanotubes with large diameters (and very small strain energies) are susceptible to a structural collapse [187, 188], it is possible that without a significant amount of strain energy the zigzag nanotubes could be geometrically *unstable*. Given some thermal fluctuations or applied strain they might collapse just like big diameter carbon nanotubes.

Finally we want to point out that the constriction of zigzag BNTs could be an important intermediate mechanism during the collapse of a zigzag BNT. It might allow for the formation of B_{12} icosahedra, which are the basic building blocks of all previously known *bulk* boron structures. The five-atomic element (see Fig. 3.21) forms part of an imaginary zigzag 6-ring, similar to the six apex atoms of a B_{12} icosahedron, as seen along each of its threefold axes. [178]

3.4.4 Summary

Generating BNTs from the BSs by a "cut–and–paste" procedure will generate *ideal* BNTs (see Sec. 3.4.1). Because the underlying two-dimensional lattice structure is rectangular rather than triangular or hexagonal, it follows that the chiral angle θ ranges from 0° to 90° ($\theta = 0°$: zigzag, $\theta = 90°$: armchair), and that chiral BNTs

3.4. NANOTUBES

do not have an axial translational symmetry. We therefore predict the existence of helical currents in ideal chiral BNTs. Furthermore we presented a band theory for ideal BNTs, employing their helical symmetry, and showed that *all* ideal BNTs are metallic, irrespective of their radius and chiral angle. BNTs could therefore be perfect nanowires, superior to carbon nanotubes.

In an independent study of armchair and zigzag BNTs we found that ideal BNTs are not the ground state of BNTs, and we identified structures of lower symmetry, which are higher in cohesive energy. The symmetries of *real* BNTs still remain to be determined, and the ideal BNTs may be seen as close approximants to real BNTs.

We also found that all BNTs, except small radius armchair types, have puckered surfaces as well as σ bonds along the armchair direction of the primitive lattice. The existence and mutual orientation of these σ bonds is crucial for our understanding of the *basic* mechanical and energetic properties of BNTs because the strain energy of the tube is mainly generated by bending those σ bonds. The multi-center bonds seem to have no effect on the strain energy. They are likely to have joint-like properties (they are easy to turn but hard to tear), which allows for a certain flexibility of these bonds, and any bonding strain could immediately be released through internal relaxations [173].

We showed that armchair BNTs, where the σ bonds lie along the circumferential direction, have rather high strain energies, whereas zigzag BNTs, where the σ bonds will lie along their axial directions, have nearly vanishing strain energies. Thus BNTs have a strain energy that depends on the nanotube's radius R as well as on the chiral angle θ: $E^{\text{B}}_{\text{strain}} = E^{\text{B}}_{\text{strain}}(R, \theta)$. We suppose that there is an individual strain energy curve for every chiral angle lying between the armchair and the zigzag curves. This is a unique property among all nanotubular materials reported so far.

The rather low strain energies in *zigzag* BNTs lead to a whole bunch of possible structural isomers, as a nanotube without any significant amount of strain energy will not be able to maintain a circular cross section. This can lead to a certain constriction of zigzag BNTs [178], and we even hypothesize that zigzag BNTs could be too unstable to really exist out in nature.

Armchair BNTs on the other hand are geometrically stabilized by their strain energies, but for armchair BNTs of rather small radii, the BNTs are unable to maintain a puckered structure necessary to align the circumferential σ bonds. In agreement with earlier studies [24, 176, 173] we expect them to flatten out and build up a smooth surface. Furthermore, we hypothesize an enhanced reactivity of small radius armchair BNTs in comparison to zigzag BNTs, which could be useful for embedding BNTs into polymers [173].

3.5 Layered Bulk Phases

So far we have shown that the boron sheet (BS) described in Sec. 3.3.3 can be used to predict the structure, the stability, the electronic, and the mechanical properties of boron nanotubes. Although it has not been observed in experiment yet, one might still wonder whether the BS has some further significance beyond its relation to boron nanotubes. Such a sheet will of course not exist as an isolated object – multiple sheets will pile up and form stacked bulk arrangements. Further questions are then: How does such a layered bulk structure look like (Sec. 3.5.2)? What is its stability in comparison with other bulk phases (Sec. 3.5.3)? Is it dynamically stable and if yes, is it responsible for the high-pressure superconductivity of elemental boron (Sec. 3.5.4)?

In this section we will connect the study of layered bulk phases with a general study of high-pressure boron and its high-pressure superconductivity.

3.5.1 Introduction: High Pressure and Superconductivity

The discovery of high-pressure superconductivity in elemental boron [9] has put boron into the focus of several experimental and theoretical groups. So far, the superconductivity is not theoretically understood. The main difficulties are that the high-pressure phases of boron are only rudimentarily known (see Sec. 3.2.2) and that the complex crystal structures complicate the theoretical treatment.

Experimental Studies

Up to now, there are only a few studies of high-pressure boron. The most prominent one is by Eremets *et al.* [9]. They performed electrical conductivity measurements under pressures up to 250 GPa and discovered that β–rhombohedral boron transforms from a semiconductor to a superconductor at $P = 160$ GPa, with $T_c = 5$ K at 175 GPa and 11.2 K at 250 GPa. The electrical resistance as function of pressure reduced almost continously from 0 to 160 GPa and stayed constant beyond 160 GPa (close to the minimum electrical conductivity). Kinks in the resistance at 30, 110, and 170 GPa could indicate phase transitions or some measurement problems.

Gerlich *et al.* [189] measured the sound velocities of the R–105 allotrope and found that the bulk and shear moduli are 205 GPa and 203 GPa, respectively. The equation of state of R–12 and R–105 boron was measured by Nelmes *et al.* [11] up to 5 and 10 GPa, respectively. They determined the the bulk modulus of β- (185 GPa) and α-rhombohedral boron (224 GPa) using neutron diffraction and x-ray diffraction, respectively. A phase transition from the β-rhombohedral (R–105) to the β-tetragonal (T-192) phase was reported by Ma *et al.* [12] at $P = 10$ GPa and $T > 2000$ K using x-ray diffraction measurements up to pressures of 30 GPa.

3.5. LAYERED BULK PHASES

They conjecture that β-tetragonal boron might be the stable phase at both high temperature and at high pressure. Furthermore they determined the bulk modulus of β-rhombohedral boron to be 205 GPa. Sanz et al. [13] determined the equation of state of β-rhombohedral boron up to $P = 100$ GPa (the bulk modulus is 210 GPa). Under hydrostatic condition they found that β-rhombohedral boron is stable up to 100 GPa and a transition to an amorphous state occurs at higher pressures. Under non-hydrostatic condition however they observed the formation of a different rhombohedral phase at lower pressures.

The partially contradictory results of different groups indicate that there is no clear picture of the high pressure phases, phase diagram, and properties of boron. Especially the very interesting superconducting phase has not been experimentally identified, so far.

Theoretical Studies

Mailhiot et al. [41] performed the first theoretical study of high-pressure phases of boron using LDA-DFT. They studied the R–12 and several closed packed structures. Among them they find R–12 $\rightarrow$ bct $\rightarrow$ fcc phase transitions at 210 and 360 GPa, respectively. This study was also the first time where metallic phases of elemental boron were proposed, as R–12 is semiconducting and the other phases are metallic. They also report the pressure induced reduction of the band gap in R–12. This effect was explored in detail by Zhao et al. [39], who found that the band gap of the R–12 phase decreases continously to zero until 140 GPa (in good agreement with the experimental onset of metalicity at 160 GPa, though it was observed in the R–105 phase) and that upon further compression the density of states (DOS) increases significantly. They found the R–12 $\rightarrow$ bct phase transition to occur at 270 GPa and no further phase transition up to 400 GPa. The discrepancy of the transition pressures can be attributed to the different exchange-correlation functionals used by Mailhiot (LDA) and Zhao (GGA).

Based on band structure calculations and the rigid-muffin-tin approximation Papaconstantopoulos et al. [42] explain the high-pressure superconductivity of boron to be based on an electron-phonon mechanism in the fcc phase. But later on it was shown by Bose et al. [43] that the fcc phase is not dynamically stable at the relevant pressures, but only at pressures exceeding ~360 GPa. They further found that bct boron is dynamically stable at lower pressures but would result in much higher T_c than what is experimentally observed and, opposite to the experimental observations, T_c would *de*crease with pressure. Thus both the fcc and the bct phases cannot account for the superconductivity in the experimentally observed range, but given the fcc or bct phases could be synthesized at high pressures, they would exhibit superconducting transition temperatures ranging from 50 to 100 K.

An important step forward towards the understanding of high-pressure boron was a study by Häussermann et al. [44] who investigated the relative stabilities of R–12, bct, fcc, and boron in the α–gallium structure (α–Ga) using DFT-GGA. They showed that α–Ga is metallic and thermodynamically more stable than bct and predict the phase transitions R–12 $\rightarrow$ α–Ga $\rightarrow$ fcc at 74 GPa and $\sim$ 800 GPa, respectively. Similar results were found by Segal and Arias [45], although they did not optimize internal parameters of the structures and therefore found α–Ga to be semiconducting. Besides Häussermann et al. show that the most stable phase of boron at different pressures is always the one that maximized the degree of sp hybridization. These findings were further elaborated by Ma et al. [46] who calculated the phonon dispersions and the electron-phonon linewidths of boron in the α–Ga structure as a function of pressure. They found that it is dynamically stable at high pressures and exhibits electron-phonon coupling to high frequency phonons. The observed increase of T_c with pressure is also consistent with their results. Hence boron in the α–Ga structure is the first real candidate to account for the high-pressure superconductivity.

A different viewpoint is brought into play by Calandra et al. [40]. They studied the dynamical properties of icosahedral $B_{12}C_2$ and found the following: if a (usually semiconducting) icosahedral network is hole doped such that the Fermi level is shifted down into the valence band, a T_c ranging from 20 to 40 K can be expected. This is caused by electron-phonon coupling to high-frequency phonons that primarily involve the icosahedral units. As the band gap of R–12, and probably also R–105, closes at high pressures [39], similar electron-phonon coupling might be expected in elemental boron. The structure that is responsible for the high-pressure superconductivity could therefore also be a common phase of elemental boron.

3.5.2 Crystal Structures and Chemical Bonding

As demonstrated above, the main difficulty in interpreting the superconducting data is closely related to the problem of determining the crystal structure. Hence before solving the problem of understanding the coupling mechanism for superconductivity, one has first to solve the problem of determining the crystal structure. As all candidate phases are still under debate and elemental boron tends to from multiple allotropes it is not unlikely that further unknown phases exist. Therefore, in the following we will study five phases: fcc, the α–rhombohedral (R–12) phase, boron in the α–Ga structure, a new Immm structure, derived from the boron sheet in Sec. 3.3, and a Fmmm structure, proposed by Boustani et al. [177]. The crystallographic data of these phases are listed in Tab. 3.5. R–12, boron in the α–Ga structure, and fcc were found in the theoretical literature to be thermodynamically stable within different pressure ranges (see above and [44, 45]). The high-pressure properties of the Immm and Fmmm phases are considered here for the first time.

3.5. LAYERED BULK PHASES

Phase	N_{atom}	Space Group	Primitive Vectors	Wyckoff Positions
fcc	1/4	Fm3m (225)	$\mathbf{a}_1 = a/2(0,1,1)$	4a: $\mathbf{B}_1 = (0,0,0)$
			$\mathbf{a}_2 = a/2(1,0,1)$	
			$\mathbf{a}_3 = a/2(1,1,0)$	
R–12	12	R$\bar{3}$m (166)	$\mathbf{a}_1 = (a/(2\sqrt{3}), -a/2, c/3)$	6h: $\mathbf{B}_1 = (x_1, x_1, z_1)$
			$\mathbf{a}_2 = (a/(2\sqrt{3}), a/2, c/3)$	6h: $\mathbf{B}_2 = (x_2, x_2, z_2)$
			$\mathbf{a}_3 = (-a/\sqrt{3}, 0, c/3)$	
α–Ga	4/8	Cmca (64)	$\mathbf{a}_1 = 1/2(a, -b, 0)$	8f: $\mathbf{B}_1 = (0, y, z)$
			$\mathbf{a}_2 = 1/2(a, b, 0)$	
			$\mathbf{a}_3 = (0, 0, c)$	
Immm	2/4	Immm (71)	$\mathbf{a}_1 = 1/2(-a, b, c)$	4e: $\mathbf{B}_1 = (0, 0, z)$
			$\mathbf{a}_2 = 1/2(a, -b, c)$	
			$\mathbf{a}_3 = 1/2(a, b, -c)$	
Fmmm	4/16	Fmmm (69)	$\mathbf{a}_1 = 1/2(0, b, c)$	8f: $\mathbf{B}_1 = 1/4(1, 1, 1)$
			$\mathbf{a}_2 = 1/2(a, 0, c)$	8g: $\mathbf{B}_2 = (1/2, 1/2, z)$
			$\mathbf{a}_3 = 1/2(a, b, 0)$	

Table 3.5: Crystal structures (phases) of elemental boron that were considered in this study; the abbreviated names for the phases are defined in Sec. 3.5.2. N_{atom} is the number of atoms per primitive (first number) and conventional (second number) unit cell. Only one number is given if both are the same. For the space groups, the number in parenthesis is defined in Ref. [117]. The Wyckoff positions are given in units of the conventional lattice vectors (for the R–12 phase the conventional cell does not exist and the Wyckoff position are defined in units of the primitive translations).

Icosahedral Phases

The icosahedral bulk phases of boron were described in detail in Sec. 3.2.2. Their fundamental building block is the B_{12} icosahedron, where each of the 12 atoms is in the preferred "inverse umbrella" bulk coordination (see Fig. 3.1(a)), i.e., it has 5 bonds within the icosahedron (intraicosahedral) and one external bond to another icosahedron (intericosahedral). The intraicosahedral bonds are primarily of three–center character, making up the triangular network of the icosahedron, whereas the external bonds are normal two–center σ bonds.[25] Within the σ bonds the charge density is more highly concentrated and more localized than within the three-center bonds. Therefore the charge density distribution $\rho(\mathbf{r})$ and the electron localization function $ELF(\mathbf{r})$ [168] can be used to detect these bonds. Furthermore the intericosahedral σ bond length is usually the shortest; e.g., our DFT–GGA calculations

---
[25] In this discussion we exclude the intericosahedral three–center bonds that are only present in α–rhombohedral boron.

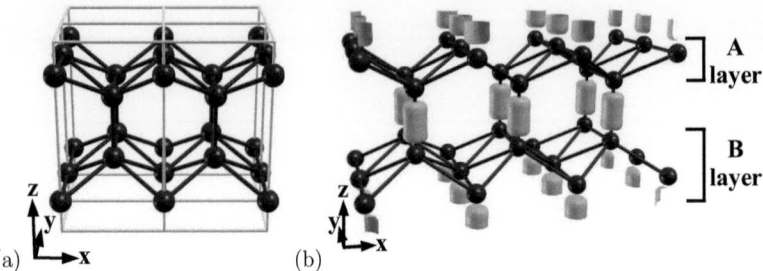

Figure 3.24: The Immm structure which is derived from the puckered boron sheet of Sec. 3.3.3. (a) Repeated conventional unit cells (green lines) illustrate the atomic arrangement. (b) Brown charge density contours at 0.9 e/Å³ show the presence of σ bonds between the two layers A and B. No σ bonds are found within the layers.

of R–12 yield (in perfect agreement with the experiment [113]) $a_{BB}^{inter} = 1.67$ and $a_{BB}^{intra} = 1.74 - 1.80$. Below we will use these three indicators to qualitatively analyze the bonding of the layered bulk structures.

Although the experimental study of Eremets et al. [9] was based on β–rhombohedral boron (R–105), we are not able to do calculations of that phase because a system size of 105 atoms per unit cell (see Sec. 3.2.2) cannot be handled by the LMTART program. Therefore we consider the R–12 phase as a representative of the icosahedral phases.

From Single Sheets to Bulk Layers

In Sec. 3.3 we discussed single boron sheets (BSs) and found that the puckered BS of Sec. 3.3.3 is a viable structure and considered it to be the boron analog of a single graphite sheet. If we transfer this model to the bulk domain, the first question is: How will multiple sheets pile up? To answer it we constructed a number of bulk models from the puckered BS and optimized their structures. This ruled out the majority of the models and only one candidate structure remained, which is shown in Fig. 3.24. It has the space group Immm (the lattice system is body–centered orthorhombic) and therefore we will call this phase *Immm* in the following. However, all of the calculations below were done in the simple orthorhombic setting of the conventional unit cell (space group Pmmm, number 47), which is also shown in Fig. 3.24. The structure can be considered as ABAB... stacking of the puckered triangular BS from Sec. 3.3.3 but there are no sp σ bonds within the layers anymore, instead one finds σ bonds of bond length $a_{BB}^{inter} = 1.71$ Å between the layers (see Fig. 3.24(b)); thus the chemical bonding in the Immm phase and in the isolated

3.5. LAYERED BULK PHASES

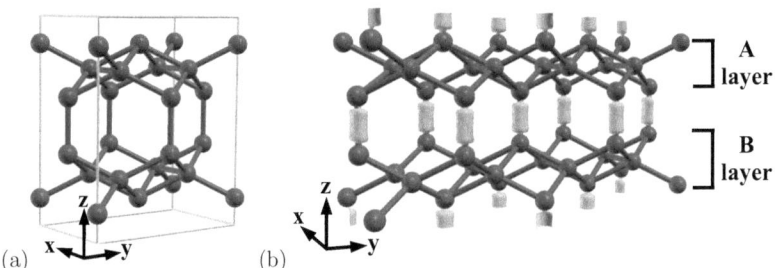

Figure 3.25: The Fmmm structure proposed by Boustani et al. [177] at $P = 0$ GPa. (a) The crystal structure in the conventional unit cell (green lines). (b) The brown features are charge density contours at 0.9 e/Å^3 that show inter-layer σ-bonds.

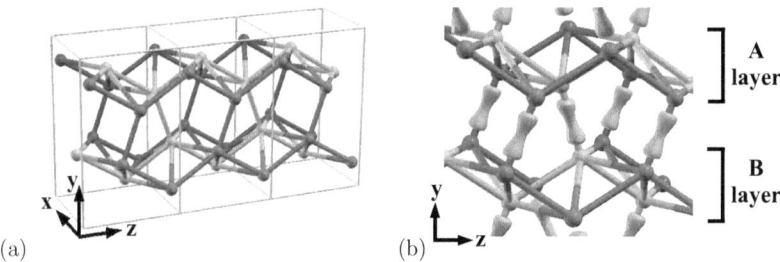

Figure 3.26: The structure of α–gallium is often visualized as 1212... stacking of planar yz-layers which contain distorted hexagons. The layers have been colored red (1), green (2), and blue (1). (a) Three conventional unit cells (green boxes) of boron in the α–gallium structure. (b) A different way of looking at the structure is to think of it as a stacking of puckered triangular xz-layers (A and B layers), bound to each other via σ bonds (shown as brown charge density contours at 0.9 e/Å^3).

BS is different. Each atom is 7-fold coordinated, having six bonds within one layer (intralayer) and one σ bond that binds to another layer (interlayer). This way each atom is close to the preferred "inverse umbrella" bulk coordination.

Another model of a layered bulk phase of elemental boron was obtained by Boustani et al. [177] by periodizing the structure of a double–layer B_{32} cluster [23]. The structurally optimized system is shown in Fig. 3.25. Its space group is Fmmm (the lattice system is face–centered orthorhombic) and again we will use this to name the structure. The unit cell contains two symmetry inequivalent atoms (see Tab. 3.5); the atoms at the 8g Wyckoff positions are four–fold intralayer and one–fold interlayer coordinated and the atoms in the 8f Wyckoff positions have six intralayer bonds only. In this atomic arrangement none of the atoms is close to the preferred "inverse umbrella" bulk coordination. Furthermore, each layer has atypical square–like "windows".[26] The charge density contours in Fig. 3.25 show that σ bonds are again only present as interlayer bonds ($a_{BB}^{inter} = 1.69$ Å).

In Fig. 3.26 we show boron in the α–Ga structure (orthorhombic base-centered). There are two ways of looking at this phase. One would be to imagine it as stacking of planar yz–layers which contain distorted hexagons. But this picture greatly neglects that the primary bonding is between atoms within puckered triangular xz–layers, where each B atom is six-fold coordinated. A 7th bond of each atom is between two of these layers and also here it is a σ bond of length $a_{BB}^{inter} = 1.74$ Å (see Fig. 3.26(b)). In this arrangement each atom is close to the preferred "inverse umbrella" bulk coordination.[27]

These three structures are very similar. We call them *layered* because all boron atoms are primarily coordinated within puckered layers (the intralayer coordination is six for most of the atoms and the interlayer coordination is one at maximum). Analysis of the charge density, the ELF, and the bond lengths (see Tab. 3.7) reveal very similar bonding for all three structures: There are three-center bonds within the layers and two–center σ bonds between them. This is akin to the bonding in the icosahedral allotropes, as described above. Thus the bonding scheme of boron bulk phases may be generalized as follows: **A three-center bonded triangular network of atoms forms basic units (icosahedra or quasiplanar layers) that are interconnected via σ bonds.** A further similarity between the icosahedral and the layered modifications is that all atoms of the α–Ga structure and the Immm phase are close to the preferred "inverse umbrella" bulk coordination. The difference is that the coordination polyhedron is a pentagonal pyramid in the case of an icosahedron

---

[26]The reader might notice the relatively huge cage–like holes that emerge between the layers. This way the boron skeleton is reminiscent of the polyhedral networks in the metal borides like MB_{12}. One might wonder whether transition metal atoms would fit into that holes and possibly stabilize the structure.

[27]It is interesting to note that the structure of α–Ga is equivalent to the one of black phosphorus, although the chemical bonding is fundamentally different.

3.5. LAYERED BULK PHASES

and a hexagonal pyramid for the layers (see Sec. 3.2.1).

Above we have qualitatively shown that the chemical bonding in the bulk layers and in the icosahedral phases is quite similar. So could layered allotropes of elemental boron be a viable possibility? To find that out we will now calculate the $T = 0$ K phase diagram and study the dynamical properties of these new structures.

3.5.3 Phase Diagram at $T = 0$ K

To judge the stabilities of the new hypothetical phases Immm and Fmmm with respect to the thermodynamically most stable phases R–12, boron in the α–Ga structure, and fcc, we calculated their enthalpies H as function of pressure P. Therefor, we first determined the energy–volume curves of the phases (see Fig. 3.27(a)). In our treatment the *volume* is the independent variable. So for a set of fixed atomic volumes we fully optimized[28] each structure and calculated the total energy. These data points were fitted to the formula

$$E(V) = c_1 + c_2 V^{-1/3} + c_3 V^{-2/3} + c_4 V^{-1}, \qquad (3.21)$$

proposed by Teter *et al.* [190]. It fits better to $E(V)$ curves that have a large volume range than the usual Birch–Murnaghan equation [191] and we see in Fig. 3.27(a) that the quality of the fit is indeed excellent. The pressure, as in Fig. 3.27(b), is then determined from

$$P(V) = -\frac{\partial E(V)}{\partial V}, \qquad (3.22)$$

and the enthalpy is given by

$$H = E + PV. \qquad (3.23)$$

In Fig. 3.27(c) we present the enthalpies of the considered phases relative to R–12. At zero temperature and at a given pressure the thermodynamically most favorable phase is the one with the lowest enthalpy. In agreement with earlier studies [44, 45] we find that with increasing pressure (up to 800 GPa) the phase transitions R–12 $\rightarrow$ α–Ga $\rightarrow$ fcc occur. The bct phase, studied by Mailhiot *et al.*, Bose *et al.*, and others [41, 43], is thermodynamically less stable than α–Ga over the whole range of pressures. That is why we do not consider the bct phase here. The transition pressures are 43 and 619 GPa, respectively. The pseudopotential DFT–GGA calculations of Häussermann *et al.* [44] obtained transition pressures of 74 GPa and 790 GPa, which qualitatively agrees with our results.

---

[28]A *full* optimization means that all lattice parameters as well as all atomic positions are optimized.

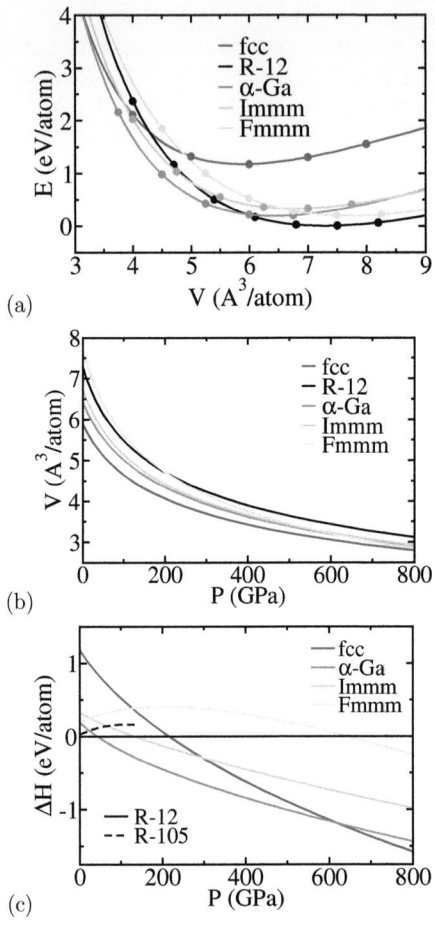

Figure 3.27: The equation of state of the five structures that were investigated. (a) Calculated total energies as function of volume. The circles are calculated data points and the lines are fits according to Eq. 3.21. (b) Atomic volume as function of pressure determined from an inversion of Eq. 3.22. (c) The enthalpies of the phases (Eq. 3.23) as function of pressure relative to R–12 (zero line). The phase with the lowest enthalpy is the most stable one. Consequently, we find the phase transitions R–12 → α–Ga → fcc at 43 GPa and 619 GPa, respectively. The dashed line is the relative enthalpy of R–105 determined from literature values of E_0 [134], V_0, B_0, B'_0 [13] (see text).

3.5. LAYERED BULK PHASES

Phase	fcc	R–12	α–Ga	Immm	Fmmm
V_0 (Å^3/atom)	5.903	7.288	6.432	6.758	7.703
E_0 (eV/atom)	1.176	0.000	0.192	0.327	0.197
B_0 (GPa)	267	214	262	230	208
B_0'	3.72	3.46	3.37	3.38	2.77

Table 3.6: The calculated equilibrium data for all structures. V_0 is the atomic volume, E_0 is the binding energy per atom relative to the energy of the R–12 phase, B_0 is the bulk modulus, and B_0' the pressure derivative of the bulk modulus. All values are taken at $P = 0$ GPa.

The equilibrium properties of the allotropes, which are the atomic volume V_0, the atomic energy E_0, the bulk modulus B_0, and the pressure derivative of the bulk modulus B_0', are given in Table 3.6. The values were extracted from the $E(V)$ fits according to equation 3.21.[29] Literature values for R–12 are $V_0 = 7.341$ Å^3/atom, $B_0 = 213$ GPa (x-ray measurements [11]), and $B_0' = 3.5$ (DFT-LDA calculations [13]) which agree very well with our results. The alpha–Ga phase was studied by Ma et al. [46] with pseudopotential calculations ($V_0 = 6.240$ Å^3/atom, $B_0 = 265$ GPa and $B_0' = 3.26$). Our FP–LMTO results, including the structural parameters given in Tab. 3.7, agree well with theirs.

An analysis of the two contributions $E(P)$ and $PV(P)$ to the enthalpy as function of pressure reveals that for $P \neq 0$ the PV term is dominating and thus primarily drives the different phase transitions.

The fcc phase has the smallest volume at all pressures (see $V(P)$ in Fig. 3.27(b)). Therefore $P \cdot V(P)$ increases the least under compression and this lowers the *relative* enthalpy more as compared to the other structures. This is clearly visible in Fig. 3.27(c) where the negative slope of fcc is the steepest of all phases. But its high atomic energy at equilibrium (1.176 eV/atom with respect to R–12, see Tab. 3.6) prevents it from being thermodynamically favorable until 619 GPa. The opposite case is R–12, which has big atomic volumes at all pressures, causing the PV term to increase the most among the phases. This quickly overcomes its low equilibrium energy and makes R–12 unfavorable at high pressures. The Fmmm structure is similar, but at high pressures it can lower its atomic volume more than the other phases. Nevertheless it possesses a rather high energy there (see Fig. 3.27(a)), so the PV term alone cannot stabilize the phase. At ambient conditions boron in the α–Ga structure and the Immm phase have relatively low energies and atomic volumes that are intermediate between R-12 and fcc. This combination causes them to have low

---

[29]The equilibrium condition $P = 0$ applied to Eqn. 3.22 defines the atomic volume V_0 and the atomic energy E_0 as the minimum of a $E(V)$ curve. The bulk modulus is given by $B_0 = -V_0(\partial P/\partial V)_{V_0}$, and the pressure derivative of the bulk modulus is $B_0' = (\partial B/\partial P)_{P=0}$.

relative enthalpies within an intermediate pressure range. So at high pressures below 600 GPa layered phases of boron are thermodynamically favorable. This is the most striking result of our phase diagram analysis.

For qualitative comparison we also show the relative enthalpy of β–rhombohedral boron (R–105) in Fig. 3.27(c). The curve was determined by inserting recent literature values of E_0 (calculated [134]), V_0, B_0, B_0' (measured [13]) into a Birch–Murnaghan equation of state [191]. As already discussed in Sec. 3.2.2 we see that R–105 is less stable than R–12 and that its instability increases further under compression. This is likely to be caused by its high atomic volume of $V_0 = 7.685$ Å^3/atom.

Overall the phase diagram can be summarized as follows: At low pressures ($P < 100$ GPa) the icosahedral phases are stable, at extreme pressures ($P > 600$ GPa) fcc is present and in the intermediate range the α–Ga structure, a layered phase of elemental boron, is thermodynamically favorable.

How do these theoretical results compare to the high–pressure experiments discussed in Section 3.5.1? First of all it is important to note that that majority of the experiments use the metastable R–105 as starting material [11, 9, 12, 13] and not R–12, which is difficult to obtain in good quality (the only study that uses R–12 is the one by Nelmes et al. [11]). In the phase diagram the R–105 curve intersects with α–Ga at 23 GPa and with Immm at 67 GPa. So below 100 GPa one could expect up to three different phase transitions. But within that pressure range, at room temperature and under hydrostatic conditions R–105 was found to be stable [13]. These experimental findings do not mean that R–12, α–Ga, and Immm are not thermodynamically more favorable, it only means that in practice R–105 is a very stable structure and the theoretical phase transitions do not necessarily take place in the experiment. To support this point further let us consider that even at ambient conditions boron does not tend to occupy the ground state structure, instead several metastable phases (as described in Sec. 3.2.2) can exists. This is the reason why the thermodynamic ground state is unknown up to now. So the pronounced polymorphism of boron makes it quite difficult to relate our theoretical phase diagram to the experiment.

Let us further note that all phases that we consider are metallic except the semiconducting R–12 and R–105. Consequently, the theoretical R–12/R–105 → α–Ga transition below 50 GPa (see Fig. 3.27(c)) would correspond to a semiconductor–metal transformation. However, experimentally such a transformation was not reported below 160 GPa [9] and there is only one theoretical transition pressure that falls into that range: It is the R–12 → Immm transformation at 134 GPa. This poses the Immm phase as a candidate to explain the superconductivity of boron under high pressure.

3.5. LAYERED BULK PHASES

Phase	α–Ga		Immm		Fmmm	
Pressure (GPa)	0	210	0	210	0	210
V/V_0	1.000	0.667	1.000	0.646	1.000	0.605
a (Å)	2.934	2.580	2.775	2.314	3.404	2.956
b (Å)	5.313	4.651	1.865	1.624	5.339	5.103
c (Å)	3.259	2.824	5.155	4.584	6.652	4.849
y	0.154	0.157	–	–	–	–
z	0.090	0.086	0.169	0.170	0.127	0.153
a_{BB}^{inter} (Å)	1.74	1.54	1.71	1.51	1.69	1.48
	1.80	1.59	1.87	1.61	1.70	1.48
a_{BB}^{intra} (Å)	1.88	1.63	1.88	1.62	1.78	1.55
	1.92	1.66	–	–	2.36	1.75

Table 3.7: Structural data for the three phases that were examined in detail at $P = 0$ and $P = 210$ GPa. The equilibrium volume V_0 is given in Tab. 3.6. a, b, c are the lattice constants and y and z are the internal parameters as defined in Tab. 3.5. a_{BB}^{inter} is the bond length between two layers (interlayer) and a_{BB}^{intra} are the bond lengths within the triangular layers (intralayer). All parameters correspond to theoretical energy minima, obtained from structural optimizations. The independent variable in our calculations is the atomic volume V (not the pressure).

3.5.4 Electronic Structure, Phononic Structure, and Superconductivity

The primary outcome of the phase diagram study in the last section is, that within a pressure range of 100 GPa $< P <$ 600 GPa the α–Ga structure, a layered phase of elemental boron, is thermodynamically favorable. This is exactly the range where superconductivity in boron was observed and that is why the layered phases are likely to be responsible for this effect.

This section is dedicated to the study of electronic and vibrational properties of α–Ga, Immm, and Fmmm in connection with superconductivity. We will consider two pressures: P=0 and P=210 GPa. The second one was chosen to lie well within the range where superconductivity was experimentally observed. Moreover, the dynamical stability of the layered structures still has to be shown before they can be considered as high–pressure allotropes of boron. The structural parameters of the three phases are given in Tab. 3.7. They were obtained by geometry optimizations as described in Sec. 3.2.4.

Icosahedral Phases

Before we study the layered allotropes let us briefly discuss results on R–12 and R–105 and their possible implications for the high–pressure superconductivity.

Measurements by Eremets et al. [9] showed that R–105 transforms from a semiconductor at P=0 to a (bad) metal at P=160 GPa; the electrical resistance continously reduces with pressure until 160 GPa and stays constant beyond that point. The continuous reduction of the resistance is consistent with the theoretical finding that the band gap in R–12 gradually reduces under compression [41, 39, 44] until it finally closes at about 140 GPa [39].[30] Our calculations on R–12 clearly confirm that trend. A similar behavior was found in R–105 [39] although no precise determination of the metalization pressure exists, yet. Calandra et al. [40] studied the dynamical properties of icosahedral $B_{12}C_2$ and found the following: if a (usually semiconducting) icosahedral network is hole doped such that the Fermi level is shifted down into the valence band, a T_c ranging from 20 to 40 K can be expected. As compression of elemental boron also leads to a metalization, similar electron-phonon coupling could be expected, if the electronic structure at the Fermi level is similar in the two cases. So there are strong indications that a common phase of elemental boron, just as R–12 or R–105, could be responsible for the high-pressure superconductivity. A study of the electron–phonon coupling in compressed R–12 or R–105 is beyond the scope of this work but should surely be addressed in the future.

The α–Gallium Structure

Boron in the the α–Ga structure was first shown to be thermodynamically favorable at high–pressures by Häussermann et al. [44] and Segal et al. [45]. It was then studied in detail by Ma et al. [46] who calculated structural, electronic, and vibrational properties, as well as the electron–phonon coupling as function of pressure with pseudopotential DFT-GGA calculations. They found that it is is always metallic, dynamically stable at high pressures, and exhibits electron-phonon coupling to high frequency phonons. The observed increase of T_c with pressure is also consistent with their work. Their results at $P = 215$ GPa compare very well with ours at $P = 210$ GPa concerning the structure (see Tab. 3.7), density of states (DOS), band structure, and phonon dispersion.[31] We also obtained good agreement for the linewidth dispersion, except along the Γ-Z direction; the discrepancies will be discussed below.

---

[30]Note that DFT usually underestimates band gaps. His shifts the theoretical metalization point to lower pressures.

[31]The $Z-T$ and $\Gamma-Y$ directions in Ma et al. [46] and our band structures and phonon dispersion plots are different because they chose the points $T = T_2$ and $Y = Y_2$ to be in the second Brillouin zone, while in our plots $T = T_1$ and $Y = Y_1$ are in the first Brillouin zone.

3.5. LAYERED BULK PHASES

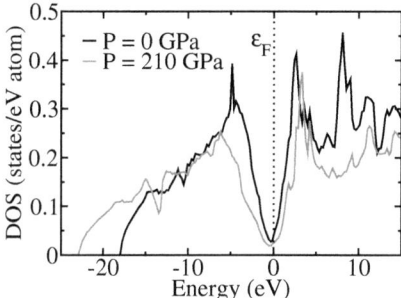

Figure 3.28: The electronic density of states (DOS) of boron in the α–Ga structure at $P = 0$ and $P = 210$ GPa. The Fermi energy ε_F is set equal to zero.

Electronic Structure The electronic density of states (DOS) and the band structures at the two pressures are given in Figs. 3.28 and 3.29. The two plots reflect the typical pressure broadening of the band structure which is caused by the reduction of the interatomic distances. But besides that there is surprisingly little pressure evolution. At the Fermi level the DOS has a global minimum, which can be an indication of structural stability. Its magnitude is reduced with pressure, i.e., $D(\varepsilon_F) = 0.040$ and 0.027 states/eV/atom at $P = 0$ and 210 GPa, respectively. The value of $D(\varepsilon_F)$ is very small indicating that boron in the α–Gallium structure is a semimetal and that its Fermi surface (FS), as shown in Fig. 3.30, has relatively small volume. This has some technical implications, as a small (and also flat in our case) FS needs a very fine k–point sampling in the Brillouin zone in order to resolve its highly localized features. Later on we will see that this issue leads to some convergence problems for the calculation of phonon linewidths. The orbital character of the band structure in Fig. 3.29 indicates that s states are primarily present in the lowest two bands, where they hybridize with p states (visible for example along the $Z - T$ or $S - R$ directions). Overall the bands are dispersed throughout the different directions of the Brillouin zone (BZ) showing that the electronic system is fully three-dimensional and that boron in the α–Gallium structure is *layered* only in a geometrical sense (see Sec. 3.5.2). At the Fermi level we primarily find p states except at 0 GPa, where along the $T - Y$ direction there is some s character right at ε_F; at 210 GPa this band is shifted to higher energies. While generally the bands are dispersive in all directions, the FS only exits in a small area around the $k_y - k_z$ plane. Similar to the whole band structure it exhibits surprisingly little pressure evolution. It consists of a big part centered around the Γ point that is hole-like and parts around the Y point and tubular features that are electron-like. To understand the results on

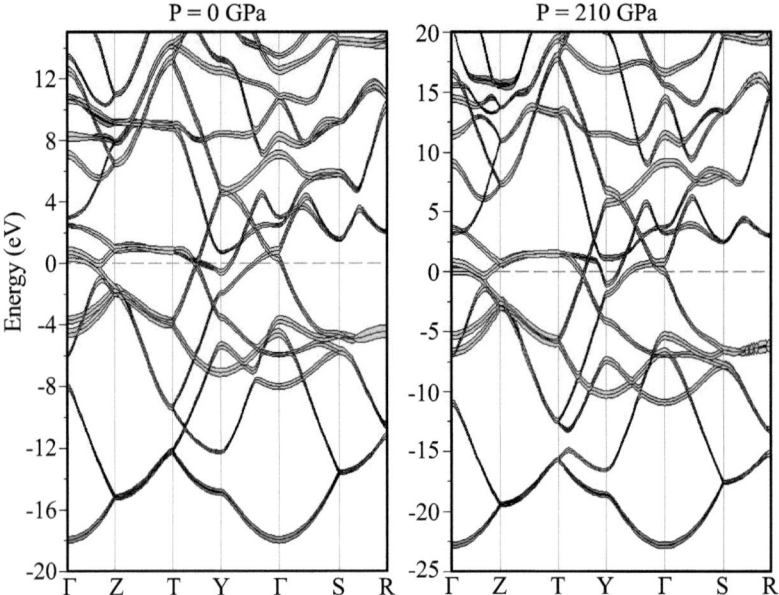

Figure 3.29: The band structures of boron in the α–Ga structure at $P = 0$ GPa and $P = 210$ GPa. The width of the bands is proportional to their orbital character; red correspond to s and green to p character. The Fermi energy is set equal to zero (blue line). The position of the special points within the first Brillouin zone is shown in Fig. 3.30.

3.5. LAYERED BULK PHASES

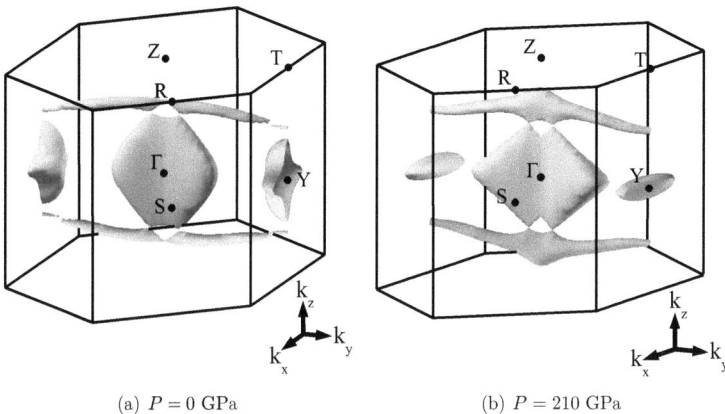

(a) $P = 0$ GPa

(b) $P = 210$ GPa

Figure 3.30: The Fermi surfaces of boron in the α–Ga structure at (a) $P = 0$ GPa and (b) $P = 210$ GPa within the first Brillouin zone (black lines). The position of the special points that are used in the band structure plots in Figs. 3.29 and 3.31 are also shown.

electron–phonon coupling, that will be discussed below, let us consider the nesting of the FS. The so called nesting function

$$\chi_\mathbf{q} = \sum_{\mathbf{k}nn'} \delta(\varepsilon_{\mathbf{k}n} - \varepsilon_\mathrm{F})\delta(\varepsilon_{\mathbf{k}+\mathbf{q}n'} - \varepsilon_\mathrm{F}) \propto \sum_{nn'} \oint_L \frac{dL_\mathbf{k}}{|\mathbf{v}_{\mathbf{k}n} \times \mathbf{v}_{\mathbf{k}+\mathbf{q}n'}|}$$

quantifies the available phase space for electron scattering across the FS (see also Eq. 2.60). It is the closed line integral over the intersections $L_\mathbf{k}$ of an undisplaced FS and one that is displaced by the vector $\mathbf{q}$; $\mathbf{v}_{\mathbf{k}n}$ is the Fermi velocity. In our case $\chi_\mathbf{q}$ will be non–zero only for $\mathbf{q}$ vectors close to and within the $q_y - q_z$ plane, because the flat shape of the FS restricts possible intersections to that region. At 210 GPa the band structure at ε_F has a flat band along Γ–Z. This is reflected in the central part of the FS being extremely thin and flat along that direction and the Fermi velocity is also lowest there (not shown). $\chi_\mathbf{q}$ will have big values for $\mathbf{q}$ vectors along Γ–Z, close to Γ, because the flat part of the displaced FS will strongly overlap with the flat part of the undisplaced one and the small and also parallel Fermi velocity vectors will further enhance $\chi_\mathbf{q}$. This effect is akin to the strong nesting of the cylindrical FS sheets of MgB$_2$ for $\mathbf{q}$ vectors along the Γ − A (q_z) direction [192, 193]. At 0 GPa this flat band is shifted up in energy (relative to the Fermi level) and the FS is also not completely flat along Γ–Z. Consequently we do not find enhanced FS nesting along

108   CHAPTER 3. NOVEL PHASES OF ELEMENTAL BORON

Γ–Z anymore. The features around the Y point, which looked like "half a disk" at 210 GPa, have a star-like morphology at 0 GPa, i.e., it is more extended along $Y-T$. The upper and lower bulges along that direction are the only parts of the FS that have s character. There will be considerable nesting between these bulges and the central part of the FS for $\mathbf{q}$ vectors close to Y, because these parts are equal in size and nearly parallel resulting in collinear Fermi velocities that make the integrand in Eq. 2.60 singular. To summarize, the FS nesting function $\chi_\mathbf{q}$ has a very sharp $\mathbf{q}$ dependence, it is non–zero only for $\mathbf{q}$ vectors close to and within the $q_y - q_z$ plane, at 0 GPa it peaks in the vicinity of the Y point, and at 210 GPa along Γ–Z, close to Γ.

Phononic Structure The phonon dispersions along directions in the $q_y - q_z$ plane are given in Fig. 3.31 for $P = 0$ and 210 GPa. At both pressures we do not find any imaginary frequencies which shows that boron in the α–Ga structure is dynamically stable over a broad pressure range. In contrast to the electronic bands the phonon dispersion exhibits a strong pressure evolution. This is also visible in phonon density of states $F(\omega)$ in the right-hand panels of Fig. 3.31 which, in addition to the usual pressure broadening, exhibits clear changes of its shape.

The phononic structure was calculated on a 4^3 and subsequently on a 6^3 q–point mesh and is relatively well represented on the 6^3 mesh, i.e., $F(\omega)$ is not smooth but the position and size of basic features are not changed when going from 4^3 to 6^3. The latter mesh was used to make Fig. 3.31, where there are only four data points along each special line in Fig. 3.31. The spline-interpolated connections between the points have been found by analyzing the phonon eigenmodes and their symmetry. The phonon symmetry labels that are given at the special points and along selected special lines in Fig. 3.31 are the q/k–point irreducible representations according to the convention of Miller and Love [194]. For example the acoustic modes at the Γ point have the symmetries Γ_3^-, Γ_2^-, and Γ_4^-, or the low frequency mode at the Y point at 0 GPa has the symmetry Y_3^-. The usual spectroscopic labels (point group irreducible representations) at the Γ point are given in Tab. 3.8 for the optical frequencies together with their Raman and infrared activity. The table indicates that many modes are Raman active and that Raman scattering could be used to experimentally identify the α–Ga structure at high pressures and to verify our calculations.

Electron–Phonon Coupling Phonon modes that couple to electrons are indicated in Fig. 3.31, where the *area* of the red circles is proportional to the mode coupling constant $\lambda_{\mathbf{q}\nu}$ (see Eq. 2.63). In the right-hand panels of the same figure the Eliashberg function $\alpha^2 F(\omega)$ (see Eq. 2.64) is plotted in red. In the discussion above we found that the nesting function $\chi_\mathbf{q}$ is non–zero only for $\mathbf{q}$ vectors close to

3.5. LAYERED BULK PHASES

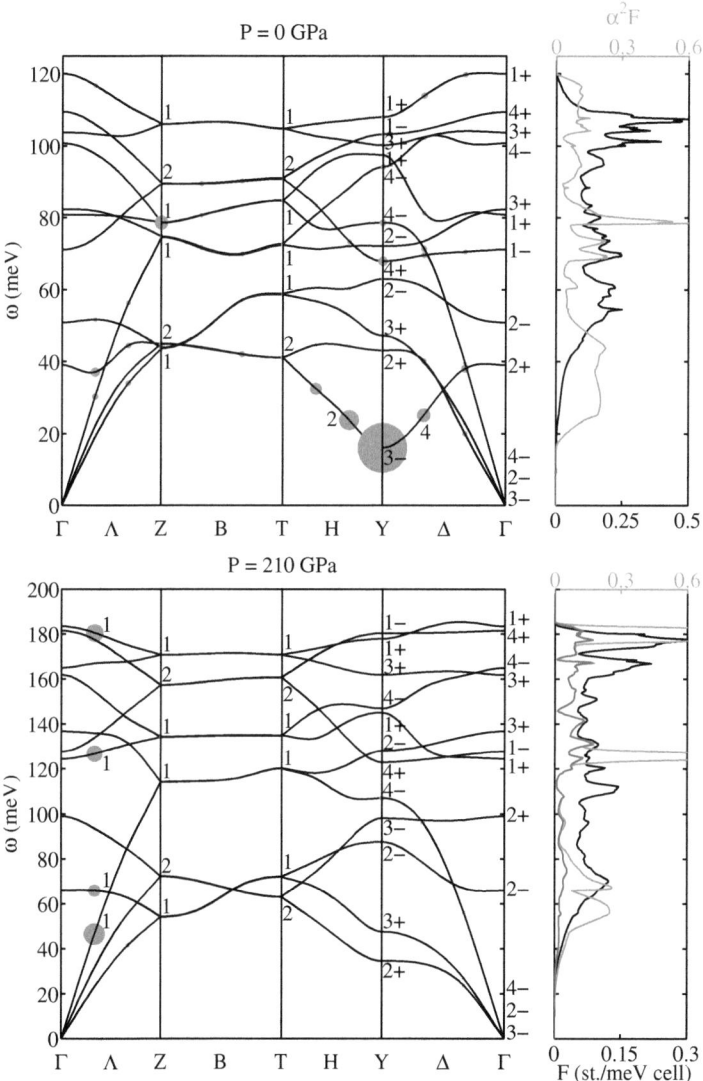

Figure 3.31: Phonons and electron–phonon coupling of boron in the α–Ga structure at 0 GPa and 210 GPa. Left–hand panels: Phonon dispersion $\omega(\mathbf{q})$. The area of the red circles is proportional to the mode coupling constant $\lambda_{\mathbf{q}\nu}$. The symmetry of the phonons is given at special points/lines. Right–hand panels: phonon density of states $F(\omega)$ (black line) and Eliashberg Function $\alpha^2 F(\omega)$ (red and blue lines).

mode	Γ_2^+	Γ_2^-	Γ_1^-	Γ_1^+	Γ_3^+	Γ_4^-	Γ_3^+	Γ_4^+	Γ_1^+
mode	B_{1g}	B_{1u}	A_u	A_g	B_{3g}	B_{2u}	B_{3g}	B_{2g}	A_g
activity	R	IR	-	R	R	IR	R	R	R
$\omega_{\Gamma\nu}(0)$	39	51	71	81	82	101	104	109	120
$\omega_{\Gamma\nu}(210)$	99	66	128	124	137	165	162	181	183

Table 3.8: The optical Γ point phonon frequencies $\omega_{\Gamma\nu}$ of boron in the α–Ga structure at 0 and 210 GPa in meV, their symmetries (first row: convention of Miller and Love [194], second row: spectroscopic labels) and spectroscopic activity (R = Raman active, IR = infrared active).

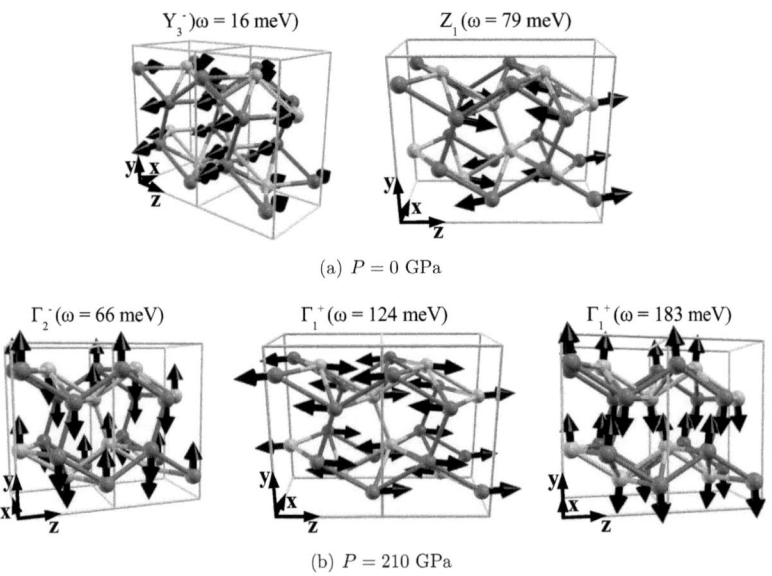

Figure 3.32: Phonon modes with strong electron–phonon coupling of boron in the α–Ga structure at (a) 0 GPa and (b) 210 GPa. The black arrows on top of the equilibrium structure indicate the atomic displacement pattern. The repeated green lined boxes are the phonon commensurate conventional supercells. For Γ and Y point phonons the cells are identical to the conventional unit cells as shown in Fig. 3.26, for the Z_1 phonon the cell is twice as big (and only one cell is shown).

3.5. LAYERED BULK PHASES

Pressure	q–mesh	$\omega_{\log}$	ω_1	ω_2	λ	T_c ($\mu^* = 0.13$)	T_c ($\mu^* = 0.20$)
0	4^3	342	440	553	1.06	22.7	15.2
	6^3	382	467	571	0.60	6.1	1.9
210	4^3	1162	1289	1393	0.22	0.0	0.0
	6^3	1039	1208	1358	0.53	9.7	1.9

Table 3.9: Parameters of the Eliashberg function $\alpha^2 F(\omega)$ and T_c of boron in the α–Ga structure evaluated on different **q**–meshes. The pressure is in GPa and T_c and the coupling–weighted phonon moments $\omega_{\log}$, ω_1, ω_2 are in K. λ is the total coupling constant and T_c is estimated using the McMillan equation (Eq. 2.67) with two different values for the Coulomb pseudopotential μ^*. All quantities are defined in Secs. 2.4.3 and 2.4.4.

and within the $q_y - q_z$ plane. This is due to the flat shape of the FS. Consequently the mode coupling constants (and the phonon linewidths) will be zero outside that region. Thus the shape of the FS gives rise to electron–phonon coupling that is confined to phonons within (or close to) the $q_y - q_z$ plane of the BZ. Therefore, in Fig. 3.31, we only plot the phonon dispersion along directions that are within that plane. Different parameters that characterize $\alpha^2 F(\omega)$, calculated on the two q–point meshes that were considered, are given in Tab. 3.9. While at the two pressures the coupling–weighted phonon moments $\omega_{\log}$, ω_1, ω_2 are converged to within $\pm 12\%$, $\pm 6\%$, and $\pm 3\%$, respectively, the total coupling constants λ change significantly and show that the convergence with respect to the q-mesh is not sufficient. The very sharp q–dependence of the FS nesting function $\chi_\mathbf{q}$ makes it necessary to sample q–space very accurately in order to describe the system correctly. This effect was encountered for many systems before, for example in compressed lithium [56, 57].

P = 210 GPa At $P = 210$ GPa the only contributions to electron–phonon coupling within the BZ come from phonons along $\Gamma - Z$ (Λ direction) and $\Gamma - Y$ (Δ direction), where $\lambda_\mathbf{q} = \sum_\nu \lambda_{\mathbf{q}\nu} \sim 1$. But most of all there is very strong coupling to all phonons of Λ_1 symmetry close to the Gamma point (the big red dots along Λ), i.e., in a very small area of q-space. This can again be traced back the the FS nesting that is very strong in that region (see discussion above). On the 6^3 q-mesh (used to make Fig. 3.31) the point $\mathbf{q} = (0, 0, 1/6)$ is the only one that falls into that region, while on the 4^3 mesh this region is not sampled at all. This explains why λ increases that strongly upon refining the q-mesh (see Tab. 3.9). The four strong peaks in $\alpha^2 F(\omega)$ all come from that single q–point. If this point is excluded, as done in the blue line in the lower right part of Fig. 3.31, the peaks disappear and the total coupling is only 0.15. This would actually be too small to induce superconductivity. In a calculation on a 12^3 q-mesh along Λ only, we found that the linewidths of the three strongly coupling optical Λ_1 modes increase the more we come close to Γ and the one of the acoustic mode decreases. The decreasing coupling of the latter mode is physically reasonable as coupling to an acoustic mode at the Γ point is not possible. In the theoretical framework that we are using, we are unfortunately not able to determine the Γ point couplings, since the nesting function is singular there and Eq. 2.61 is ill-defined,[32] but it is very likely that electron–phonon coupling is strongest there. At the zone center the three strongly coupling optical Λ_1 phonons become two Raman active Γ_1^+ (A_g) modes and one infrared active Γ_2^- (B_{1u}) mode (see Tab. 3.8). Their displacement patterns are shown in Fig. 3.32(b). All modes correspond to intralayer distortions: The y elongations in Γ_2^- leave the interplanar distance fixed and the low and high frequency Γ_1^+ modes corresponds to distortions along z and y, respectively.

We conclude that any significant contributions to electron–phonon coupling in boron in the α–Ga structure come from Γ point and Λ phonons that are close to Γ. In order to make quantitative statements about superconductivity this very small region of q–space must be sampled very carefully by q–meshes much finer than 6^3.

The linewidths that we obtained in the strongly coupling region have extreme values: At $\mathbf{q} = (0, 0, 1/6)$ and $(0, 0, 1/12)$ the linewidths of the two high frequency Λ_1 phonons are $\gamma_6 = 25$ and 36 meV and $\gamma_{12} = 58$ and 117 meV, respectively. As these values are about $20-60\%$ of the actual phonon frequency, it would imply that the nearly free particle picture for the phonons breaks down. This is quite unlikely. Therefore we checked the convergence with respect to the k-mesh representing the electronic structure of the equilibrium system in the linear response calculations (used to determine the electron-phonon matrix elements and the FS nesting function). All calculations on boron in the α–Ga structure were done on a 48^3 k–mesh, which is the very best we could technically do with LMTART. On that mesh the linewidths are usually converged better than 15%, which is not accurate but acceptable, but they are not at all converged in the region of very strong coupling, where the relative error reaches up to 75%. We conclude that the extreme linewidths are numerical artefacts of an improper representation of the electronic structure that is unable to resolve the very strong FS nesting along Λ. Thus in order to make reliable statements about electron–phonon coupling in boron in the α–Ga structure it is not only necessary to increase the precision in q–space but also in k–space. As these calculations are very expensive, we have to postpone them to the future. However, the numerical problems that we encounter here are not new and the slow convergence of the FS nesting function with respect to the k–point sampling for regions of very strong nesting is a problem that might also be overcome by analytical treatments (see Kong et al. [192] or Dolgov et al. [193]).

Finally, let us compare our results at $P = 210$ GPa with the ones by Ma et al. [46]

---

[32]We checked that the usual Γ point divergence of the nesting function $\lim_{\mathbf{q} \to 0} \chi_\mathbf{q} = \infty$ (see Sec. 2.4.3) is not responsible for the increase of the linewidths.

3.5. LAYERED BULK PHASES

at 215 GPa. As mentioned above, we obtained excellent agreement for the structural, electronic and phononic properties but we have some disagreement concerning the results on electron–phonon coupling. First, in the Eliashberg functions only the positions of the two high–frequency peaks agree, whereas the overall shapes are quite different. Second, the order of magnitude of our linewidths at $P = 210$ GPa compares well with the values Ma *et al.* give at 160 GPa, but we find very strong differences along $\Gamma - Z$. The source of the disagreement is clearly that our linewidths, determined by using the unbiased tetrahedron method, are not converged along that direction, but it is either not clear whether their linewidths are converged (which were calculated with a smearing method). Third, at the zone center the low frequency optical phonon mode that couples strongly is Γ_2^- (B_{1u}) in our calculations, while Ma *et al.* find that it is Γ_2^+ (B_{1g}). Furthermore, they only use a $4 \times 3 \times 4$ q-mesh, which we showed to be too coarse to provide reasonable results for $\alpha^2 F(\omega)$ and λ.

To summarize, neither the study of Ma *et al.* nor our study can decide whether boron in the α–Ga structure is responsible for the measured $T_c = 9$ K superconductivity of elemental boron at $P = 210$ GPa. This is due to the sharp q–dependence of the FS nesting function that requires very fine sampling of the phonon and electron BZs and none of this was achieved so far. Nevertheless, the qualitative picture of Ma *et al.* can be verified by our analysis and the values for T_c in Tab. 3.9 show that the strong electron–phonon coupling to Γ point and Λ phonons (close to Γ) should lead to a measurable superconductivity.

P = 0 GPa It is very interesting that boron in the α–Ga structure is dynamically stable even at $P = 0$. As its cohesive energy is 0.19 eV higher than the one of R–12 or 0.16 eV higher than R–105 we predict that this phase is stable at ambient conditions.[33] Furthermore, Fig. 3.31 reveals that there are much more strongly coupling phonons than at 210 GPa and within the $q_y - q_z$ plane almost every phonon has $\lambda_\mathbf{q} \geq 1$. This explains why we find much more softened[34] phonon branches at $P = 0$ than at 210 GPa and why the pressure evolution of the phononic structure is relatively strong. Due to the disappearance of the flat electronic band at ε_F along the $\Gamma - Z$ direction, the FS nesting is not extreme anymore and the coupling there is as strong as in the rest of the $q_y - q_z$ plane.

The system exhibits large coupling to a strongly softened optical branch (H_2, Y_3^-, Δ_4 in Fig. 3.31) in an area close to the Y–point, which is highest at Y and decreases the more we get away from Y. This again can be ascribed to the FS nesting as discussed above. This effect generates the broad plateau in $\alpha^2 F(\omega)$ in the low

---

[33] Mind that the difference in cohesive energy between the two forms of carbon, graphite and diamond, is 0.2 eV [195], i.e., it is even bigger than here.

[34] The frequency of so called *softened* phonons is reduced due to strong electron–phonon interactions. This phonon softening is a manifestation of the interaction–induced renormalization of the quasiparticle energies (phonon frequencies).

frequency range between 20 and 50 meV. At the Y point $\omega_{Y_3^-} = 16$ meV, $\gamma_{Y_3^-} = 2.2$ meV, and $\lambda_{Y_3^-} = 33.3$, i.e., the linewidth is about 14% of the frequency. However, all linewidths are smaller than 4 meV and we assume that they are converged with respect to the k–point sampling (again, we used a 48^3 k–mesh). As λ scales with $1/\omega^2$ (see Eq. 2.63), the small phonon frequency is the mathematical reason why lambda is so big. The atomic displacement pattern of the Y_3^- mode is shown in Fig. 3.32(a). Again it corresponds to intralayer distortions that leave the interplanar distance fixed. It is interesting that the very strong coupling (and softening) to the Y_3^- phonon is reduced with pressure. At 210 GPa no significant coupling exists anymore.

The peak in $\alpha^2 F(\omega)$ at $\omega = 79$ meV is caused by two degenerate Z_1 phonons with $\lambda_{Z_1} = 2.5$ and $\gamma_{Z_1} = 3.9$ meV. The latter is the biggest linewidth at $P = 0$ and $\gamma_{Z_1}/\omega_{Z_1}$ is only 5%. Figure 3.32(a) shows the atomic elongations that correspond to these Z_1 modes. They are again intralayer distortions, the interlayer distance is not fixed now and primarily undergoes bond rotations.

At both pressures we have seen that the phonon modes that couple strongly to electrons are associated with distortions that do not necessarily involve stretching of the interlayer σ bonds, e.g., Y_3^- and Γ_2^- leave it fixed. We therefore do not find any significant similarities between the mechanism of electron–phonon coupling in boron in the α–Ga structure and in MgB$_2$.

Overall we observe a drastic change in electron–phonon coupling at the two considered pressures. This drastic change is also visible in the two Eliashberg functions $\alpha^2 F(\omega)$ in Fig. 3.31, whose shapes are totally different. While at $P = 210$ the coupling is confined to a very small region in q–space (which requires very accurate BZ sampling), at 0 GPa the coupling is strong in the whole $q_y - q_z$ plane and varies weakly with **q**. We therefore believe that the q-mesh of 6^3 is representative for this system and the the values given in Tab. 3.9 are good estimates for λ and T_c. The relatively big jump in λ when changing from 4^3 to 6^3 q sampling is caused by overweighting the contributions of the Y point on the 4^3-mesh. To summarize, we predict that boron the α–Ga structure is likely to exist at ambient conditions where it would be a $T_c = 2$ to 6 K conventional superconductor.

The Immm Structure

All calculations of the Immm structure were done in the simple orthorhombic setting of the conventional unit cell (space group Pmmm).

The electronic densities of states (DOS) and band structures at $P = 0$ and 210 GPa are shown in Fig. 3.33 and 3.34, respectively. Besides the broadening of the bands we again observe relatively little pressure evolution in the electronic structure. As in boron in the α–Ga structure, the states at ε_F primarily have p character but there are more states at the Fermi level now, i.e., $D(\varepsilon_F) = 0.137$ and 0.097

3.5. LAYERED BULK PHASES

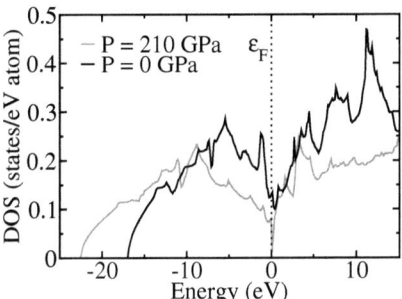

Figure 3.33: The electronic density of states (DOS) of the Immm structure at $P = 0$ and $P = 210$ GPa. The Fermi energy ε_F is set equal to zero.

states/eV/atom at $P = 0$ and 210 GPa, respectively. And again the bands are dispersive along the different directions of the Brillouin zone (BZ) showing that the electronic system is fully three-dimensional and that the Immm structure is *layered* only in a geometrical sense (see Sec. 3.5.2). At $P = 0$ there is one band along $\Gamma - Y$ and along $S - X_1$ that touches ε_F in a hole-like fashion. At 210 GPa the dispersion of four bands is totally flat near ε_F and along $S - R$. Two bands have p character and are exactly at the Fermi level, while the other two have s character and are slightly higher in energy. These flat bands cause the DOS to jump at ε_F. Such a peak can be sign of a structural instability. And indeed a look at the phonon dispersion in Fig. 3.35 shows that at 210 GPa one (transverse) acoustic phonon brach along $\Gamma - Y$ is imaginary.[35] Thus the Immm structure is dynamically unstable at that pressure. At 0 GPa we only find a *single* imaginary frequency in the phonon BZ. It is a transverse acoustic mode at the point $\mathbf{q} = (1/8, 0, 0)$ along $\Gamma - X$. This point-like instability could indicate that a superstructure, modulated by the vector $\mathbf{q} = (1/8, 0, 0)$, is lower in energy (more stable) than the current structure. Such a system would contain 32 atoms per conventional unit cell. Comparing the two pressures we see that different phonon branches are imaginary. This strongly suggest that at an intermediate pressure there will be no imaginary frequencies anymore and the Immm structure might be fully stable. Besides the usual pressure hardening the phonon dispersions exhibit a very strong pressure evolution. This is clearly visible in the phonon DOS in the right–hand part of Fig. 3.35 whose overall shape is very different at the two pressures and the phononic band gap at $P = 0$ and about 100 meV is closed at the second pressure. The phonon DOS also indicates that the

---

[35]On the $8 \times 6 \times 4$ q–mesh only the points $\mathbf{q} = (0, 1/6, 0)$, $(0, 1/3, 0)$, and $(1/8, 1/3, 0)$ have *one* imaginary frequency.

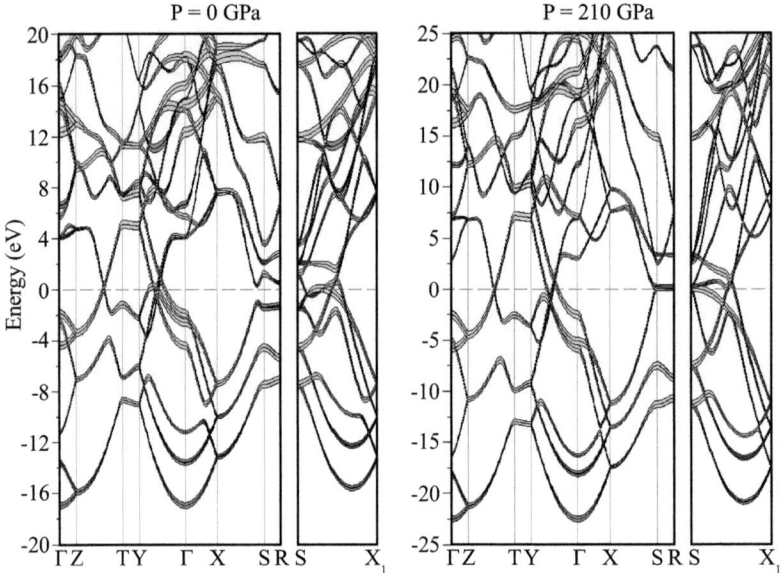

Figure 3.34: The band structures of the Immm structure at $P = 0$ GPa and $P = 210$ GPa. The width of the bands is proportional to their orbital character; red correspond to s and green to p character. The Fermi energy is set equal to zero (blue line). The position of the special points within the first Brillouin zone is shown in Fig. 3.36.

3.5. LAYERED BULK PHASES

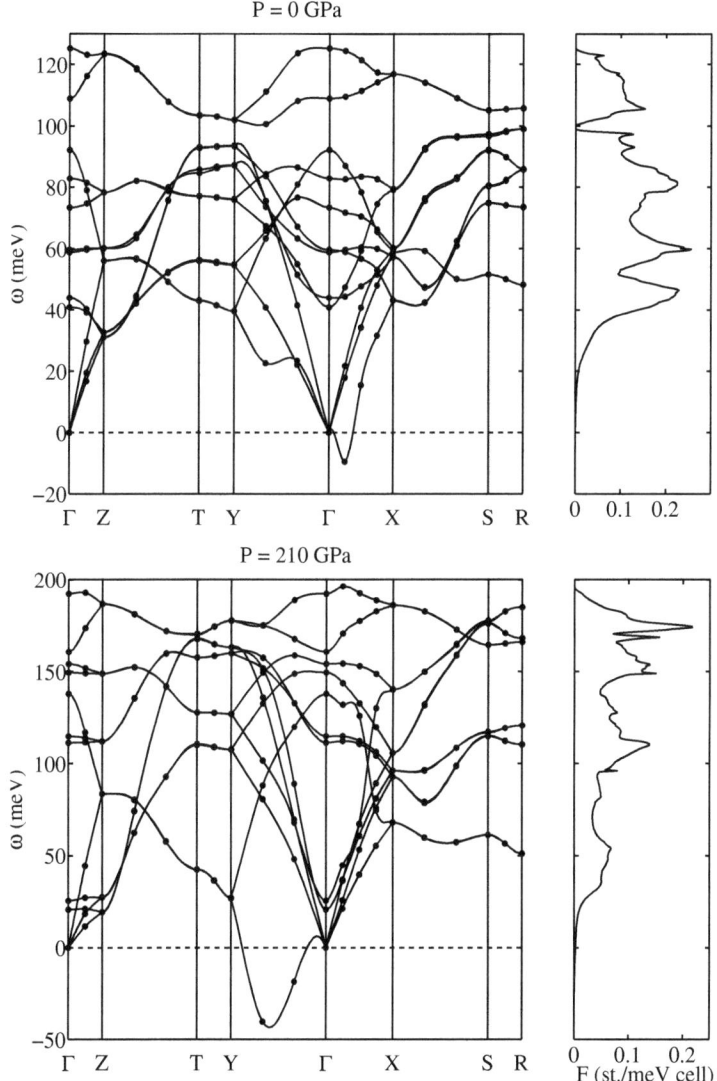

Figure 3.35: Phonons of boron in the Immm structure at 0 GPa and 210 GPa. Left–hand panels: Phonon dispersion $\omega(\mathbf{q})$ along lines of symmetry. Negative frequencies represent imaginary frequencies. Right–hand panels: phonon density of states $F(\omega)$.

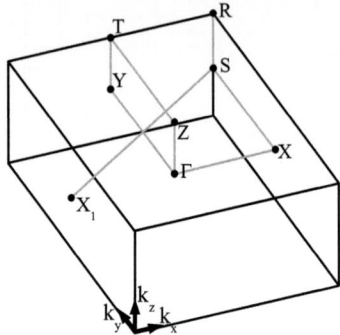

Figure 3.36: The fist Brillouin zone of the Immm structure and the position of the special points (black dots) and the selected directions (orange) for the band structure plots in Figs. 3.34 and 3.35.

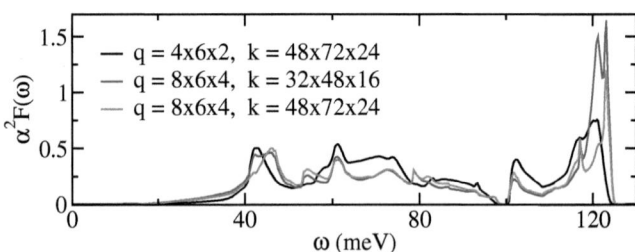

Figure 3.37: The Eliashberg function $\alpha^2 F(\omega)$ of boron in the Immm structure at 0 GPa, evaluated on different q and k meshes.

3.5. LAYERED BULK PHASES

number of imaginary modes is very small in both cases. We conclude that the Immm structure appears to be close to a stable structure and further investigations have to be done.

The phononic structure was calculated on a $4 \times 6 \times 2$ and subsequently on a $8 \times 6 \times 4$ q–point mesh and is well represented on the latter one, i.e., the functions $F(\omega)$ evaluated on both meshes agree qualitatively. The finer mesh was used to generate Fig. 3.35 and the spline-interpolated connections between the individual data points (black dots) have been found by analyzing the phonon eigenmodes.

Since the Immm structure is dynamically unstable at the two pressures, we do not study the electron–phonon coupling in detail but rather present general results of the $P = 0$ case. The Eliashberg functions in Fig. 3.37 were evaluated on different q–point and k–point meshes. The k–meshes represent the electronic structure of the unperturbed system in the linear response calculations, this includes the all–important Fermi surface and the nesting function. In general, the convergence of $\alpha^2 F(\omega)$ is reasonable, except in the high–frequency end where the different numerical parameters lead to varying results. The strong electron–phonon coupling to high–frequency modes involves relatively big phonon linewidths reaching up to 21 meV. But again we face the problem that the linewidths are not converged with respect to the k–point sampling. When changing the k–mesh from $32 \times 48 \times 16$ to $48 \times 72 \times 24$ individual linewidths were corrected by up to several hundred percent and only very few modes were converged better than 15%. However the BZ averaging seems to cancel out many of these huge errors and, except for the high–frequency range, $\alpha^2 F(\omega)$ seems to be reasonably converged. The values in Tab. 3.10 are calculated from $\alpha^2 F(\omega)$ and show that a superconducting transition temperature of 7 to 16 K can be expected. The above described numerical inaccuracies are not likely to dramatically alter this fundamental result.

To summarize, boron in the Immm structure is dynamically unstable at $P = 0$ and 210 GPa. But as different phonon breaches are imaginary in the two cases, we expect the phase to be stable at intermediate pressures. At $P = 0$ we only find a single imaginary frequency throughout the BZ. This could indicate that a modulated superstructure is actually stable. Ignoring the instability at ambient conditions the system would be a 7 to 16 K conventional superconductor. The present results are already quite promising and and further investigations on the Immm phase should be done.

The Fmmm Structure

The electron and phonon DOS[36] of boron in the Fmmm structure are given in Fig. 3.38. The system would be metallic both at 0 and 210 GPa but a look at

---
[36]The phonon DOS was evaluated on a 5^3 q–mesh.

q-mesh	$\omega_{\log}$	ω_1	ω_2	λ	T_c ($\mu^* = 0.13$)	T_c ($\mu^* = 0.20$)
$4 \times 6 \times 2$	764	821	877	0.70	20.5	9.0
$8 \times 6 \times 4$	677	755	827	0.68	16.2	6.6

Table 3.10: Parameters of the Eliashberg function $\alpha^2 F(\omega)$ and T_c of the Immm structure at $P = 0$ GPa, evaluated on different q–meshes (the k–mesh is $48 \times 72 \times 24$). T_c and the coupling–weighted phonon moments $\omega_{\log}$, ω_1, ω_2 are in K. λ is the total coupling constant and T_c is estimated using the McMillan equation (Eq. 2.67) with two different values for the Coulomb pseudopotential μ^*. All quantities are defined in Secs. 2.4.3 and 2.4.4.

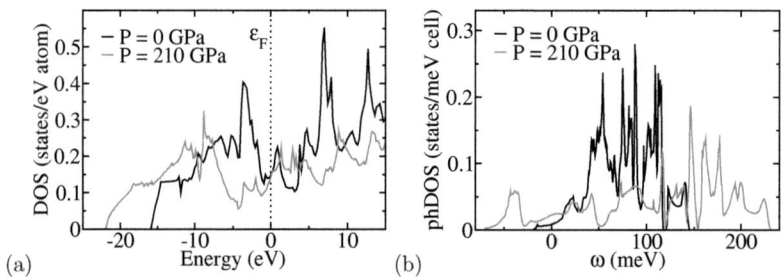

Figure 3.38: The (a) electronic and (b) phononic density of states (DOS) of boron in the Fmmm structure at $P = 0$ and $P = 210$ GPa. The Fermi energy ε_F in (a) is set to zero. Negative frequencies in (b) represent imaginary frequencies.

3.5. LAYERED BULK PHASES

the phonons shows that the phase is dynamically unstable as there are a significant number of imaginary frequencies. This strong instability at both pressures indicates that the Fmmm structure can be ruled out as a possible allotrope of elemental boron at ambient and at high pressures.

Its instability can be explained by a violation of a basic structural rule for elemental boron. That is, the Fmmm phase is the only one of the three layered phases under consideration where the atoms are not close to the preferred "inverse umbrella" bulk coordination. This was discussed in detail in Sec. 3.5.2. Furthermore, Immm is thermodynamically unfavored as shown in Sec. 3.5.3.

3.5.5 Summary

In section 3.5 we studied different bulk phases of elemental boron: fcc, α–rhombohedral, α–Ga, Immm, and Fmmm. All of them are metallic, except α–rhombohedral boron which is a semiconductor at ambient conditions. The phases α–Ga, Immm, and Fmmm were studied in detail. We called them "layered" because the boron atoms are primarily coordinated within quasiplanar layers and have at most one bond connecting two of these layers. The Immm phase is a simple ABAB... stacking of the boron sheet studied in Sec. 3.3.3 and Fmmm was proposed by Boustani et al. [177]. Both systems can be seen as an extension of the Boustani Aufbau principle (introduced in Sec. 3.2.1) to the bulk domain.

Within the quasiplanar layers the atoms are held together by three–center bonds, and between them by two–center σ bonds. We showed that boron can reach its preferred "inverse umbrella" bulk coordination in the α–Ga and Immm structures where the basic bonding is akin to the one in the icosahedral phases. This allows to define the following generalized picture of the chemical bonding in boron solids: A three-center bonded triangular network of boron atoms forms basic units (icosahedra or quasiplanar layers) that are interconnected via σ bonds.

A calculation of the $T = 0$ K phase diagram of the five considered phases showed that the layered structures α–Ga and Immm are thermodynamically favorable at pressures between 100 and 600 GPa. This is exactly the range where high–pressure superconductivity was experimentally observed. Below 100 GPa the common icosahedral phases are favored and above 600 GPa closed packed systems are stable. In agreement with earlier studies we found that with increasing pressure (up to 800 GPa) the phase R–12, α–Ga, and fcc are thermodynamically most favored, with theoretical phase transitions occurring at 43 and 619 GPa, respectively.

Because of their potential to explain the high–pressure superconductivity of elemental boron the electronic and phononic structure as well as the electron–phonon coupling of the layered phases were studied in detail.

The electronic band structures of boron in the α–Ga structure and the Immm phase exhibit very little pressure evolution, except the usual pressure broadening.

The states at the Fermi surfaces primarily have p–orbital character and the bands are fully three-dimensional, which shows that both structures are "layered" only in a geometrical sense.

Boron in the α–Ga structure is dynamically stable at the two considered pressures $P = 0$ and 210 GPa. As its Fermi surface only exits in a small area around the k_y–k_z plane of the electron Brillouin zone electron–phonon coupling is restricted to phonons within the $q_y - q_z$ plane of the phonon Brillouin zone. At 210 GPa any significant contributions to electron–phonon coupling come from a very small region of q–space, i.e., the Γ point and the $\Gamma - Z$ directions close to Γ. But neither this nor an earlier study [46] can decide whether boron in the α–Ga structure is responsible for the measured $T_c = 9$ K superconductivity of elemental boron at $P = 210$ GPa. This is due to the sharp q–dependence of the Fermi surface nesting function that requires very fine sampling of the phonon and electron Brillouin zones and none of this was achieved so far. Nevertheless, both studies show that the strong electron–phonon coupling at 210 GPa should lead to a measurable superconductivity. At $P = 0$ GPa we found strong electron–phonon coupling within the whole $q_y - q_z$ plane of the phonon Brillouin zone and particularly strong coupling to a Y_3^- optical phonon (at the Y point). In contrast to the 210 GPa case the numerical parameters seem to be converged and we predict that boron in the α–Ga structure could not only be stable at ambient conditions but also a $T_c = 2$ to 6 K conventional superconductor. Finally, our study showed that the electron–phonon coupling at the two considered pressures is drastically different.

Boron in the Immm structure is not dynamically stable at $P = 0$ and 210 GPa. But as different phonon branches are imaginary in the two cases, we expect the phase to be stable at intermediate pressures. At $P = 0$ we only find a single imaginary frequency throughout the BZ. This could indicate that a modulated superstructure is actually stable. Ignoring the instability at ambient conditions the system would be a 7 to 16 K conventional superconductor. The present results are already quite promising and and further investigations on the Immm phase should be done.

The Fmmm structure it thermodynamically unfavored and dynamically unstable. It can therefore be ruled out as a possible allotrope of elemental boron at ambient and high pressures. Its instability can be explained by an unfavorable coordination of the boron atoms and its relatively big atomic volume.

3.6 Summary and Conclusions

This chapter was dedicated to the study of novel phases of elemental boron. Our approach was based on the Boustani Aufbau principle that was introduced in Sec. 3.2. The Aufbau principle is very general scheme that predicts several classes of novel boron materials going far beyond the known icosahedral phases. Sections 3.3 and

3.6. SUMMARY AND CONCLUSIONS

3.4 studied boron sheets and boron nanotubes and Sec. 3.5 attempted to extend the ideas of Aufbau principle to the bulk domain.

First, we examined a number of different structure models for broad boron sheets (BSs) in Sec. 3.3. All of them are metallic, and we found that for a 16 atom supercell, the model with a simple up–and–down puckering is be the most stable one. Its chemical bonding can be put into the following preliminary picture: on the one hand the sheet is held together by homogeneous multi-center bonds, on the other hand there are linear sp hybridized σ bonds exclusively lying along the armchair direction of the sheet. The rather anisotropic bond properties of the sheets lead to different elastic moduli C_x and C_y for stretching the BS in the x and in the y direction. Furthermore puckering of the BS may be understood as a key mechanism to stabilize the sp σ bonds. Our results indicate that the sheet analyzed in this study is the boron analog of a single graphene sheet, a possible precursor of boron nanotubes (BNTs).

Constructing BNTs from the BSs by a "cut–and–paste" procedure will generate *ideal* BNTs. Because the underlying two-dimensional lattice structure is rectangular rather than triangular or hexagonal, it follows that the chiral angle θ ranges from $0°$ to $90°$ ($\theta = 0°$: zigzag, $\theta = 90°$: armchair), and that chiral BNTs do not have an axial translational symmetry. We therefore predict the existence of helical currents in ideal chiral BNTs. Furthermore we presented a band theory for ideal BNTs, employing their helical symmetry, and showed that *all* ideal BNTs are metallic, irrespective of their radius and chiral angle. BNTs could therefore be perfect nanowires, superior to carbon nanotubes.

In an independent study of armchair and zigzag BNTs we found that ideal BNTs do not represent the ground state of BNTs, and we identified structures of lower symmetry, which are higher in cohesive energy. The symmetries of *real* BNTs still remain to be determined, and the ideal BNTs may be seen as rather close approximants to real BNTs. We also found that all BNTs, except small radius armchair types, have puckered surfaces, and σ bonds along the armchair direction of the primitive lattice. The existence and mutual orientation of these σ bonds is crucial to understand the *basic* mechanical and energetic properties of BNTs because the strain energy of the tube is mainly generated by bending those σ bonds. The multi-center bonds seem to have no real effect on the strain energy. They are likely to have joint-like properties (they are easy to turn but hard to tear), which allows for a certain flexibility of these bonds, and any bonding strain could immediately be released through internal relaxations. We showed that armchair BNTs, where the σ bonds lie along the circumferential direction, will have rather high strain energies, whereas zigzag BNTs, where the σ bonds will lie along their axial directions, will have nearly vanishing strain energies. Thus BNTs have a strain energy that depends on the nanotube's radius R as well as on the chiral angle θ: $E_{strain}^{B} = E_{strain}^{B}(R, \theta)$. We suppose that there is an individual strain energy curve for every chiral angle lying between the armchair and the zigzag curves. This is a unique property among all nanotubular materials reported so far. The rather low strain energies in *zigzag* BNTs lead to a whole bunch of possible structural isomers, as a nanotube without any significant amount of strain energy will not be able to maintain a circular cross section. This can lead to a certain constriction of zigzag BNTs, and we even hypothesize that zigzag BNTs could be too unstable to really exist out in nature. *Armchair* BNTs on the other hand are geometrically stabilized by their strain energies, but for armchair BNTs of rather small radii, the BNTs are unable to maintain a puckered structure necessary to align the circumferential σ bonds. In agreement with earlier studies we expect them to flatten out and build up a smooth surface. Furthermore, we hypothesize an enhanced reactivity of small radius armchair BNTs in comparison to zigzag BNTs, which could be useful for embedding BNTs into polymers.

The above findings define a consistent picture of boron sheets and boron nanotubes and unify former studies on these materials into a generalized theory. However, while writing up this thesis the field of boron nanomaterials has evolved further. Szwacki et al. [33] proposed a model for a particularly stable spherical cluster (boron fullerene) and based on our theory and this fullerene model Tang et al. [196] and Yang et al. [197] proposed improved models for boron sheets and boron nanotubes. Overall the field has received considerable attention in the media as recent articles in scientific newspapers [198, 199, 200, 201, 202, 203] reveal.

In Sec. 3.5 we extended the ideas of the Boustani Aufbau principle to the bulk domain and asked whether layered bulk phases, similar to graphite, may also exist for boron. Our study then tried to approach the following questions: What do such layered bulk structures look like? What is their stability in comparison with other bulk phases? Are they dynamically stable and if yes, are they responsible for the high–pressure superconductivity of elemental boron?

In order to approach these question we studied different bulk phases: fcc, α–rhombohedral, α–gallium, Immm, and Fmmm. All of them are metallic, except α–rhombohedral boron which is a semiconductor at ambient conditions. The phases α–Ga, Immm, and Fmmm were studied in detail. We called them "layered" because the boron atoms are primarily coordinated within quasiplanar layers and have at most one bond connecting two of these layers. The Immm phase is a simple ABAB... stacking of the broad boron sheet described above and Fmmm was proposed by Boustani et al.

Within the quasiplanar layers the atoms are held together by three–center bonds, and between them by two–center σ bonds. We showed that boron can reach its preferred "inverse umbrella" bulk coordination in the α–Ga and Immm structures where the basic bonding is akin to the one in the icosahedral phases. This allowed to define the following generalized picture of the chemical bonding in boron solids: A three-center bonded triangular network of boron atoms forms basic units (icosahedra or quasiplanar layers) that are interconnected via σ bonds.

3.6. SUMMARY AND CONCLUSIONS

A calculation of the $T = 0$ K phase diagram of the five considered phases showed that the layered α–Ga structure is thermodynamically favorable at pressures between 100 and 600 GPa. This is exactly the range where high–pressure superconductivity was experimentally observed. Below 100 GPa the common icosahedral phases are favored and above 600 GPa fcc boron is stable. In agreement with earlier studies we found that with increasing pressure (up to 800 GPa) the phase R–12, α–Ga, and fcc are thermodynamically most favored, with theoretical phase transitions occurring at 43 and 619 GPa, respectively.

Because of their potential to explain the high–pressure superconductivity of elemental boron the electronic and phononic structure as well as the electron–phonon coupling of the layered phases were studied in detail.

The states at the Fermi surfaces primarily have p–orbital character and the electronic band structures of boron in the α–Ga structure and the Immm phase are fully three-dimensional, which shows that both structures are "layered" only in a geometrical sense.

Boron in the α–Ga structure is dynamically stable at the two considered pressures $P = 0$ and 210 GPa. As its Fermi surface only exits in a small area around the k_y–k_z plane of the electron Brillouin zone electron–phonon coupling is restricted to phonons within the $q_y - q_z$ plane of the phonon Brillouin zone. At 210 GPa any significant contributions to electron–phonon coupling come from a very small region of q–space, i.e., the Γ point and the $\Gamma - Z$ directions close to Γ. But neither this nor an earlier study [46] can decide whether boron in the α–Ga structure is responsible for the measured $T_c = 9$ K superconductivity of elemental boron at $P = 210$ GPa. This is due to the sharp q–dependence of the Fermi surface nesting function that requires very fine sampling of the phonon and electron Brillouin zones and none of this was achieved so far. Nevertheless, both studies show that the strong electron–phonon coupling at 210 GPa should lead to a measurable superconductivity. At $P = 0$ GPa we found strong electron–phonon coupling within the whole $q_y - q_z$ plane of the phonon Brillouin zone and particularly strong coupling to a Y_3^- optical phonon (at the Y point). In contrast to the 210 GPa case the numerical parameters seem to be converged and we predict that boron in the α–Ga structure is not only stable at ambient conditions but also a $T_c = 2$ to 6 K conventional superconductor.

Boron in the Immm structure is dynamically unstable at $P = 0$ and 210 GPa. But we expect the phase to be stable at intermediate pressures. Stability might also be achieved at $P = 0$ by a modulated superstructure. At ambient conditions the system would be a 7 to 16 K conventional superconductor if we ignore the slight instability. The present results are already quite promising and and further investigations on the Immm phase should be done.

The Fmmm structure is thermodynamically unfavored and dynamically unstable. It can therefore be ruled out as a possible allotrope of elemental boron at ambient and high pressures.

We conclude that novel metallic bulk phases of boron, different from the known icosahedral phases, are likely to exist at elevated pressures or even at ambient conditions. Furthermore, there are strong indications that these phases are conventional superconductors with considerable high superconducting transition temperatures. However, the present results are not able to unravel the origin of the experimentally reported high–pressure superconductivity in elemental boron.

3.7 Outlook

Our findings in Secs. 3.3 and 3.4 led to a consistent picture of boron sheets and boron nanotubes and unified former studies on these materials into a generalized theory. At present the field of boron nanomaterials is evolving rapidly. Our results on the bulk systems in Sec. 3.5 only partially allow to draw a generalized picture and raise some further questions.

- The electronic structure and the chemical bonding of the elemental phases of boron should be analyzed in more detail. Multi–center bonding is a very interesting phenomenon which was hardly ever studied in bulk systems and simple orbital–based descriptions are needed. Furthermore, the subtle interplay between two–center and three–center bonding in elemental boron, which leads to complex crystal structures and novel nanostructures, lacks proper understanding and poses many questions to theory.

- There are several indications that a common phase of elemental boron, just as R–12 or R–105, could also be responsible for the observed high-pressure superconductivity. Therefore, studying electron–phonon coupling in compressed icosahedral phases should be addressed.

- In order to put the yet qualitative results for boron in the α–Ga structure at high–pressure to a quantitative level the q–point sampling should be refined.

- The slow convergence of the FS nesting function with respect to the k–point sampling for regions of very strong nesting, appears to be a severe problem. We encountered this for the Immm and the α–Ga structure in the form of unconverged electron–phonon linewidths. As the morphology of the FS is not simple, it is not clear whether analytical treatments are viable here.

- The pressure dependence of the electron–phonon coupling and the superconducting transition temperature should be studied in more detail by considering more pressure points (atomic volumes) for all systems.

3.7. OUTLOOK

- The Immm structure is dynamically unstable at $P = 0$ and 210 GPa. But it appears to be close to points of stability. Therefore, a superstructure or intermediate pressures should be considered to stabilize the phase. Furthermore, the geometrical connections between the very stable α–Ga and the Immm structure could be studied.

Chapter 4
Structure Control of Nanotubes

The success of future nanotechnologies will strongly depend on our ability to control the structure of materials on the atomic scale. For carbon nanotubes it turns out that one of their structural parameters – the chirality – may not be controlled during synthesis. In this chapter we explain the basic reason for this defect and show that novel classes of nanotubes, which are related to sheets with anisotropic in-plane mechanical properties (e.g. boron nanotubes, see Chapter 3), could actually overcome these problems. Our results further suggest that extended searches for nanotubular materials similar to pure boron might allow for one of the simplest and most direct ways to achieve structure control within nanotechnology.

4.1 Introduction

Carbon nanotubes (CNTs) [2] are certainly the most prominent member of a whole family of nanotubular materials with technologically interesting properties like WS_2 [204], BN [205, 206] or the recently discovered pure boron nanotubes [24, 25, 21]. In general the electronic and mechanical properties of single walled nanotubes depend quite strongly on their structure, which may be characterized by two parameters: the radius R and the chiral angle θ (chirality). One usually encodes R and θ by two integers (n, m) referring to the basis vectors of the underlying primitive lattice [29]. Unfortunately, it turns out that for the standard synthesis of CNTs one may achieve some control over their radii [31, 30, 32, 207], but little control over their chiralities, which implies that in general, there is little control over the properties of the end products of the synthesis. As CNTs may be either metallic or semiconducting, depending on their radii and chiralities [29], this poor structure control will also imply a rather poor control over the electronic properties of CNTs. And for nanotubular systems other than carbon, we are not necessarily facing a better situation.

4.2 Strain Energy

The strain energy of a nanotube is defined as

$$E_{\text{strain}} = E_{\text{sheet}} - E_{\text{NT}}(R, \theta), \qquad (4.1)$$

where E_{sheet} and E_{NT} are the cohesive energies [75] of the sheet and the tube, respectively. E_{strain} can be understood in the following ways: it quantifies

1. the difference in cohesive/total energy among different (R, θ) nanotubes,

2. the deformation (curvature) energy per atom, which is necessary to roll up a single sheet into a nanotube of certain radius R and chiral angle θ, and

3. it is is a measure of the mechanical tension of a nanotube. This tension stabilizes the tubular shape (it makes the tube round), and it is also responsible for radial breathing mode vibrations.

4.2.1 Carbon Nanotubes

For CNTs the strain energy refers to a graphene sheet and, as shown in Fig. 4.1(a) and discussed in textbooks [208, 29], the strain energy effectively depends on the radius R, but not on the chirality: $E_{\text{strain}} = E_{\text{strain}}(R) = C/R^2$ (in Fig. 4.1(a) $C = 2.178$ eV Å^2). This radial dependence is easy to understand: the radius is just a measure for the curvature of a CNT, and the smaller the radius is the more energy is needed to bent a graphene sheet. But why is the strain energy independent of chirality? This behavior may be attributed to the nearly isotropic in-plane mechanical properties of the graphene sheet, as quantified by its elastic moduli for stretching and bending. For example, the elastic constants C_{11} and C_{22} are the same, due to a hexagonal symmetry of the honeycomb lattice [29]. Therefore, when stretching a graphene sheet along different in-plane lattice directions, one will observe the same stiffness. From a chemical point of view this mechanical isotropy is caused by a hexagonal network of stiff sp^2 σ bonds, as shown in Fig. 4.2(a). Thus when rolling up a graphene sheet along different in-plane directions (chiral angles) to form various nanotubes with similar radii, this process will require similar deformation energies. Therefore E_{strain} will be independent of the chirality of the CNT. This mechanical behavior is analogous to a simple sheet of paper that is rolled up to form a tube. This process will require little energy for big radii, and it is becoming more and more costly with decreasing radii. But due to the isotropic in-plane mechanical properties of the paper sheet, the energy needed to roll up a paper tube is independent of the roll up direction (chiral angle). A similar behavior is also known for BN, BC$_3$ [209], or MoS$_2$ [210] nanotubes.

4.2. STRAIN ENERGY

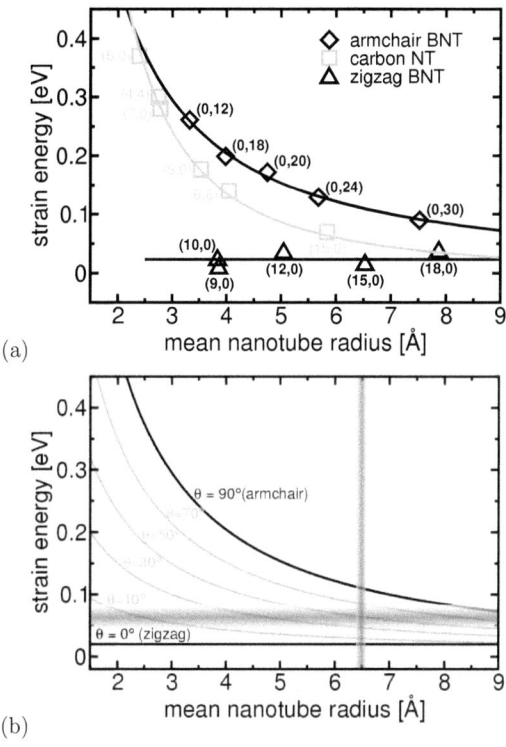

Figure 4.1: Calculated strain energies for different (n, m) carbon nanotubes (carbon NT) and (k, l) boron nanotubes (BNT) (see Chapter 3). (a) The strain energies of carbon nanotubes with different chiral angles can be described by a single curve (orange/grey), which is just a function of the tubular radius. Armchair $(0, m)$ and zigzag $(n, 0)$ boron nanotubes instead have two distinct strain energy curves (black). (b) The black curves for armchair and zigzag boron nanotubes are taken from (a), while the orange (grey) curves illustrate other chiral angles θ. The blue, horizontal line shows that the reaction conditions of the synthesis define a certain energy range for the resulting nanotubes, and the vertical line indicates how, e.g., template materials limit the radii of the nanotubes. Where both lines intersect nanotubes with similar radii and chiralities could be synthesized (see text).

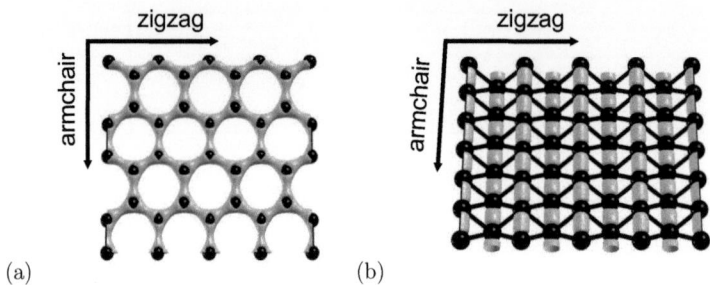

Figure 4.2: Sigma bonds in (a) graphite and (b) boron sheets. The black balls and sticks in (a) and (b) represent the atomic lattice. The orange (grey) charge density contours at (a) 1.8 and (b) 0.9 e/Å^3 show the presence of sp^2 and sp type σ bonds, respectively.

4.2.2 Boron Nanotubes

In Chapter 3 we found that the lattice structure of a boron sheet is rectangular rather than hexagonal, and therefore the chiral angle θ of BNTs ranges from 0° to 90°, in contrast to 0° to 30° known for CNTs. Thus BNTs and CNTs relate to reference lattices of different symmetry, and therefore one has to use different chiral indices for CNTs and BNTs. In the following we use (n, m) for CNTs and (k, l) for BNTs. Furthermore, the boron sheet has anisotropic in-plane mechanical properties, where the ratio between the elastic constants C_{11} and C_{22} was calculated to be $C_{22}/C_{11} \sim 2$, (see Chapter 3: $C_{11} = 420$ GPa, $C_{22} = 870$ GPa). Fig. 4.2(b) shows that the boron sheet has parallel linear chains of stiff sp σ bonds lying along the sheet's armchair direction, whereas along the zigzag direction, one finds softer bonds of three center character [211] (not shown in Fig. Fig. 4.2(b)). Therefore, stretching the sheet along its armchair direction (involving C_{22}) will be much harder than stretching it along its zigzag direction (involving C_{11}). Similarly, bending the sheet along the armchair direction, which involves bending of rather stiff σ bonds, will take more energy (strain energy) than bending the BNTs along their zigzag direction, where no σ bonds will be affected. This is quite evident from Fig. 4.1(a), where we show that armchair BNTs have high strain energies, whereas zigzag BNTs have nearly vanishing strain energies. Thus the strain energies of BNTs depend on their radii and their chiral angles: $E_{\text{strain}} = E_{\text{strain}}(R, \theta)$. For the whole range of chiral angles (0° $< \theta <$ 90°) we expect individual strain energy curves located between both extremes, as illustrated in Fig. 4.1(b). Thus the boron sheet will basically behave like a piece of cloth that

4.2. STRAIN ENERGY 133

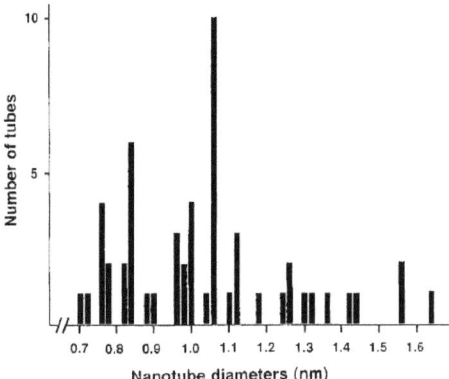

Figure 4.3: The distribution of nanotube radii as reported by Iijima et al. [30]. The distribution forms a well peaked bell-shaped curve that can be shifted or broadened by the specific reaction conditions of the synthesis.

is reinforced along one direction with parallel chains of stiffeners (the σ bonds). Bending the cloth along the lines of stiffeners (armchair direction) takes significantly more energy than bending the cloth perpendicular to it (zigzag direction).

In Chapter 3 we further predicted that BNTs are always metallic, independent of their radii and chiralies. The Fermi surface of the boron sheet has some well pronounced contours in the 2D Brillouin zone (see Fig. 3.14), and backfolding of the Fermi surface into the 1D Brillouin zone of a BNT is possible for any radius and any chirality. For graphene, on the other hand, the Fermi surface just exists at the K points of the Brillouin zone, and backfolding of these special points into the 1D Brillouin zone of a (n,m) CNT is possible, but only if $(n-m)$ turns out to be a multiple of 3 [29]. Thus for CNTs their electronic properties (semiconducting versus metallic) vary quite strongly with radius and chiral angle, but their energies are independent of chirality. BNTs are just the opposite, in the sense that their electronic properties will not depend on the structure type, but their total energies actually do.

4.3 Structure Control

The spectrum of nanotube radii obtained during a synthesis of CNTs (see Fig.4.3) will depend on the specific reaction conditions (temperature, pressure, catalyst, reaction gas, etc.), and it can be shifted and/or broadened by changing these conditions. Nevertheless, the CNT chiralities remain random and rather uncontrollable. This was demonstrated by Iijima *et al.*, Bethune *et al.*, and Journet *et al.* [30, 31, 32], who were all synthesizing single-walled CNTs using the arc-discharge method, but due to different reaction conditions, they reported different mean diameters of 1.0, 1.2, and 1.4 nm, respectively. Furthermore they note that the chiral angles vary quite strongly for a given tube diameter. These observations can be explained if we assume that the reaction conditions of the synthesis primarily influence the total energy of the synthesized nanotubes. In other words the reaction conditions will determine a certain energy range for the resulting nanotubes, and by virtue of the $E = E(R)$ dependence of CNTs (see Fig. 4.1(a)) this energy range fixes a certain range of radii, but leaves the chirality totally unspecified. In contrast to this, the energies of nanotubes like BNTs, which are derived from a sheet with anisotropic in-plane mechanical properties, will strongly depend on their chiralities and radii $E = E(R, \theta)$, and the reaction conditions will influence both structural parameters. Such a behavior might ultimately allow for better structure control among nanotubular materials, because now the different chiral angles should be energetically separable and thus experimentally accessible.

As for the radii of the nanotubes, the former may be controlled by growing the nanotubes out of porous materials with well defined pore sizes [21, 207]. And their energies (strain energies) may be controlled by tuning the reaction conditions. Thus after limiting the ranges of radii and strain energies, it should be possible to actually synthesize a rather narrow range of nanotubes with similar radii and chiralities (see Fig. 4.1(b)), or even one specific type of nanotube, only.

4.4 Summary

In summary, we have proposed a different route to achieve control over the structure of nanotubes. By analyzing the unfavorable case of CNTs, we have shown that the current inability to control the chirality of nanotubes is caused by isotropic in-plane mechanical properties of the related graphene sheets, leading to isoenergetic nanotubes with similar radius, but a whole range of different chiral angles. This "degeneracy" is lifted for nanotubes that are derived from a reference sheet with anisotropic in-plane mechanical properties. As demonstrated for the case of BNTs, this anisotropy will make the different chiral angles energetically separable, and this should be experimentally accessible. Generally speaking, it might actually

4.4. SUMMARY

pay to supplement current efforts to achieve a higher degree of structural control over nanotubular materials by a systematic search for nanotubular systems, which are related to sheets with anisotropic in-plane mechanical properties. Short segments of such materials may then serve as templates to handle less controllable materials such as CNTs, where the template will impress radius and chirality through stable intramolecular heterojunctions, as shown in [212].

Chapter 5

The Enatom Method

5.1 Introduction

The idea to describe condensed matter as a collection of (generalized) atoms dates back to to the ancient Greek atomists[1] and remains a powerful concept in modern solid state physics, reflected in site-based models, atomic orbital computational approaches, and a resurgence of interest in Wannier functions. A different starting point for the description of solids is the homogeneous electron gas, where the atomic nature is ignored at first and brought in later to account for the reality of a real solid. These are the two traditional viewpoints of condensed matter physics.

The first example how the total density $n(\bm{r})$ and the total electronic potential $v(\bm{r})$ of a solid can be separated into atomic contributions was provided by the *neutral pseudoatom* concept of Ziman [214]. In the limit of weak pseudopotentials, this approach describes a nearly–free–electron solid by overlapping atomic contributions. It is however strongly restricted by the limitation to weak pseudopotentials, which applies only to the alkali metals and may not give satisfactory results even there. The *auxiliary neutral atom* by Dagens is defined on the basis of a change in density (with zero net charge) that is induced by a screened potential in jellium [215]. This approach was inspired by density functional theory and solved numerically for Li and Na, but the author did not address the overlapping of such entities. In the work of Streitenberger a *generalized pseudoatom model* is introduced which extends the pseudoatom concept of Ziman to inhomogeneous electron–ion systems for simple metals [216]. The model is based on linear–response theory in the density functional framework and it is applied to a metal surface.

---

[1] Often ascribed to Democritus, ca. 430 B.C. See, for example [213]

5.1.1 Theoretical Background

Three decades ago Ball introduced a *generalized pseudoatom* density decomposition concept that applies to any solid [50, 51]. It is this specification that we follow in this chapter. The development of the broad pseudoatom concept and the terminology (pseudoatom; auxiliary neutral atom; generalized pseudoatom; quasi-atom) has a long history, and the term pseudoatom also means 'atom described by a pseudopotential' and 'phantom atom to tie off dangling bonds', as well as many other applications, as a literature search will readily reveal. It will therefore be useful to introduce unambiguous language in the following: instead of using the *generalized pseudoatom* terminology of Ball we introduce the term *enatom*.[2] It should be clarified that Ball's pseudoatom has nothing to do with any pseudopotential (a common use of the term).

For any reference position of atoms, the (vector) first-order change in charge density upon displacing one atom at $\boldsymbol{R}_j$ from its equilibrium position $\boldsymbol{R}_j^o$, i.e., the linear response to displacement, can be separated into its irrotational and divergenceless components

$$\frac{\partial n(\boldsymbol{r})}{\partial \boldsymbol{R}_j} \equiv \nabla_j n(\boldsymbol{r}) \tag{5.1}$$
$$= -\nabla \rho_j(\boldsymbol{r} - \boldsymbol{R}_j^o) + \nabla \times \boldsymbol{B}_j(\boldsymbol{r} - \boldsymbol{R}_j^o);$$

here $n(\boldsymbol{r})$ is the charge density of the system. An immediate result is a pair of remarkable sum rules [50]. (i) The lattice sum of the *rigid density*[3] $\rho_j(\boldsymbol{r} - \boldsymbol{R}_j^o)$ gives an exact decomposition of the crystal charge density into atomic contributions:

$$\sum_j \rho_j(\boldsymbol{r} - \boldsymbol{R}_j^o) = n(\boldsymbol{r}). \tag{5.2}$$

(ii) The lattice sum of the *deformation density* (or "backflow") $\nabla \times \boldsymbol{B}_j(\boldsymbol{r} - \boldsymbol{R}_j^o)$ vanishes identically:

$$\sum_j \nabla \times \boldsymbol{B}_j(\boldsymbol{r} - \boldsymbol{R}_j^o) = \boldsymbol{0}. \tag{5.3}$$

This strong constraint reflects that this nonrigid density is a cooperative effect of neighboring atoms, which nevertheless can be broken down uniquely into individual atomic contributions. Clearly atoms that are equivalent by symmetry have identical ρ_j and $\boldsymbol{B}_j$; an elemental solid with one atom per primitive cell only has one of each.

---

[2] Greek: 'en' denotes within or inside; 'atom' denotes indivisible part. Therefore 'enatom' connotes the indivisible part inside a system.

[3] Our definition of the rigid part ρ is predominantly positive [like $n(\boldsymbol{r})$] and differs in sign from the convention of Ball [50].

5.1. INTRODUCTION

This cooperative origin (a solid state effect) of the deformation density can be understood by considering that an atom, embedded in a jellium background, would have no deformation density. In fact, if the density is $n(\boldsymbol{r})$ when the atom is at the origin, then by translational symmetry the density is $n(\boldsymbol{r} - \boldsymbol{R}_j)$ when the atom is located at $\boldsymbol{R}_j$. Therefore $\nabla_j n(\boldsymbol{r}) = -\nabla \rho_j(\boldsymbol{r})$ and there is no nonrigid contribution. Furthermore, within standard treatments (such as the local density approximation) $\rho_j(\boldsymbol{r})$ is spherical, because the atom is embedded in an isotropic environment. Thus the deformation density and the anisotropy of the rigid density arise solely from the inhomogeneity of the system, i.e., by neighboring atoms. The other extreme is represented by a covalent solid, which is held together by strong directional bonds. The rigid density will be highly non-spherical, and the deformation density will be comparatively large, reflecting the fact that when an atom moves its bonding is disrupted.

To first order in displacements from the reference point $\delta \boldsymbol{R}_j = \boldsymbol{R}_j - \boldsymbol{R}_j^o$ (typically the equilibrium lattice), the density is given by [50]

$$n(\boldsymbol{r}; \{\boldsymbol{R}_j\}) = \sum_j [\rho_j(\boldsymbol{r} - \boldsymbol{R}_j^o - \delta \boldsymbol{R}_j) \quad (5.4)$$
$$+ \delta \boldsymbol{R}_j \cdot \nabla \times \boldsymbol{B}_j(\boldsymbol{r} - \boldsymbol{R}_j^o)].$$

The quantity inside the sum is the *enatom* of atom j and moves rigidly with the nucleus to first order. (The $\vec{R}_j^o$ in the argument of $\vec{B}_j$ can be replaced by $\vec{R}_j$ without changing the expression to first order.) Analogous decomposition and sum rules apply to the potential $v(\boldsymbol{r})$ [217]:

$$v(\boldsymbol{r}; \{\boldsymbol{R}_j\}) = \sum_j [V_j(\boldsymbol{r} - \boldsymbol{R}_j^o - \delta \boldsymbol{R}_j) \quad (5.5)$$
$$+ \delta \boldsymbol{R}_j \cdot \nabla \times \boldsymbol{W}_j(\boldsymbol{r} - \boldsymbol{R}_j^o)].$$

Since $\rho, \boldsymbol{B}$ and $V, \boldsymbol{W}$ are first order quantities, the changes in density and potential[4] can be related by linear response theory [217]. The deformation arises only from off-diagonal components of the dielectric matrix $\varepsilon(\boldsymbol{q} + \boldsymbol{G}, \boldsymbol{q} + \boldsymbol{G}')$, i.e., deviation of $\varepsilon(\boldsymbol{r}, \boldsymbol{r}')$ from the $\varepsilon(\mathbf{r} - \mathbf{r}')$ form.

Equation (5.4) allows for a transparent interpretation of the quantities ρ_j and $\nabla \times \boldsymbol{B}_j$. The total charge density $n(\boldsymbol{r}; \{\boldsymbol{R}_j\})$ of a system of *displaced* atoms is constructed from the charge densities $\rho_j(\boldsymbol{r} - \boldsymbol{R}_j^o - \delta \boldsymbol{R}_j)$ that move *rigidly* with the atoms upon displacement, plus a second part $\nabla \times \boldsymbol{B}_j(\boldsymbol{r} - \boldsymbol{R}_j^o)$ that describes how the charge density *deforms* due to nuclear displacement.

---

[4]The fields $\boldsymbol{B}$ and $\boldsymbol{W}$ are only defined up to a gauge transformation: $\boldsymbol{B} \to \boldsymbol{B} + \nabla \phi$ leads to the same physical deformation density for any scalar 'potential' ϕ. Nevertheless, it makes sense to specify $\boldsymbol{B}$ and $\boldsymbol{W}$ uniquely (up to a constant) by requiring it to be divergenceless. The Helmholtz prescription provides this divergenceless field [217].

It is important to keep in mind that, although the rigid enatom density (potential) is a specified decomposition of the crystal analog, it does not arise simply from screening of the pseudopotential (which is the case in weak pseudopotential theory). It is intrinsically a dynamically determined quantity, involving only linear response. Specifically, the enatom potential arises from a screened displaced (pseudo)potential

$$\nabla_j v = \varepsilon^{-1} \nabla_j v_{ps}, \quad (5.6)$$

while the enatom density arises from the linear change in wave functions (see also Eq. 2.37)

$$\nabla_j n = 2 \left(2Re \sum_{kn}^{occ} \psi_{kn}^* \nabla_j \psi_{kn} \right). \quad (5.7)$$

which can be obtained from first-order perturbation theory. (The first factor of 2 is for spin.)

A related quantity is the atomic deformation potential $\nabla_j \epsilon_{kn}$ (change in any band energy due to displacement of the atom at $\vec{R}_j$), given from perturbation theory by

$$\nabla_j \epsilon_{kn} = \langle kn | \nabla_j v | kn \rangle \quad (5.8)$$

in terms of enatom quantities. Khan and Allen showed the relation of this *deformation potential* to electron-phonon matrix elements [218]. Resurgent interest in electron-phonon coupled superconductivity has led to the suggestion by Moussa and Cohen that this quantity may provide insight into strong coupling [219].

Although Ball was the first to introduce the enatom decomposition and begin to make use of it, the importance of $\nabla_j n(r)$ had been recognized earlier. Sham emphasized its essence in the formulation of lattice dynamics, related it to the shift in potential by the density response function [220], and related its integral to effective charges. Its application to ionic insulators was extended by Martin, who showed that the enatom dipole and quadrupole moments are the fundamental atomic entities that underlie piezoelectricity [221].

As powerful as the enatom concept is (see discussion below), very little use has been made of it. Falter and collaborators have adopted a related *quasi-ion* idea for sublattices of multiatom compounds, and used linear response theory (or models) to evaluate sublattice charges for Si [222, 223, 224]. Ball and Srivastava calculated some aspects of the rigid and deformation parts of the density in Ge and GaAs from bond-stretch distortions [225, 226]. No calculation of single enatom quantities yet exist for any material.

5.1.2 Physical Motivation

The enatom concept will be particularly important in studying and understanding phonons and electron-phonon coupling, which requires only information arising from

5.1. INTRODUCTION

an infinitesimal displacement of atoms (thus, linear response). Current implementations of linear response theory calculate the first-order change in potential due to a given phonon, and use periodicity and Bloch's theorem to reduce the calculation to a unit cell, which is still time-consuming. This linear response problem must be solved separately for each phonon momentum q. Using the enatom concept and linear superposition, it is necessary only to calculate the enatom density and potential *once* (for each inequivalent atom in the primitive cell) and perform elementary integrals requiring only linearly superimposed, overlapping enatom potentials to calculate the phonon frequencies. (We are concerned here only with metals; Ball has shown that insulators with long-range potentials require extra considerations [51].) The electron-phonon matrix elements are even easier, as they can be reduced to calculating the matrix elements of the enatom potential of each inequivalent atom only. This might not be quite as easy as it sounds, because the enatom may in some cases have to be calculated out to a distance of several shells of neighbors to obtain convergence of the integrals. We postpone the phonon problem to future work.

For a first detailed application of this enatom concept to enhance understanding of bonding and electron-phonon coupling, we choose the simple metals Li and Al. Lithium has attracted interest due to the recent discovery that, in spite of being a simple free-electron-like metal that is not superconducting above 100 μK at ambient pressure [52], it displays high $T_c \approx$ 15-17 K in the 30-40 GPa range [6, 7, 8], and T_c = 20 K has been reported [6] around 50 GPa. This discovery made Li the best superconducting elemental metal (now equaled by yttrium [227] and apparently surpassed by calcium [228]). The evolution of the electronic structure and electron-ion scattering within the rigid muffin-tin approximation is well studied [229, 230]. Application of microscopic superconductivity theory, with phonon frequencies and electron-phonon matrix elements calculated using linear response methods [55, 56, 57, 58], has established that this remarkable T_c results from strong increase of electron-phonon coupling under pressure. The one aspect of the electron-phonon behavior in Li that is not yet understood [57] is the strong branch dependence of electron-phonon matrix elements. Application of enatom techniques promises to be an ideal way to approach the remaining questions.

Aluminum is the simplest trivalent metal, with T_c = 1.2 K at ambient pressure; superconductivity is suppressed with pressure, with T_c < 0.1 K at 6 GPa [53]. Under pressure the electronic structure of Al remains that of a free electron-like metal, and a structural transition to a hcp phase takes place only at P > 217 GPa [54]. Li, on the other hand, becomes more and more covalent and undergoes several phase transitions [5]. In the case of metals we use the term "covalency" in a loose sense to indicate the appearance of directional bonds. The vibrational and electronic properties of the two systems also display important differences. While the electronic structure, Fermi surface and vibrational spectrum of Al follow a completely normal trend (i.e., the band dispersion becomes steeper, the Fermi surface is virtually unchanged and the

	P	V/V_0	a	n_0	$N(0)$	l_{TF}	E_F	k_F
Li	0	1.00	7.98	0.79	3.41	1.13	0.27	0.52
	35[a]	0.52	6.41	1.52	2.58	1.02	0.30	0.55
	50	0.44	6.05	1.81	2.39	0.98	0.29	0.54
Al	0	1.00	7.50	2.85	2.61	0.91	0.83	0.91
	35	0.77	6.89	3.67	1.96	0.87	0.97	0.98
	50	0.73	6.75	3.90	1.85	0.86	1.01	1.00

Table 5.1: Structural and electronic properties of fcc Li and Al as a function of pressure. Except for the calculated pressure (P), which is in GPa, all the quantities are expressed in atomic units. V_0 is the theoretical equilibrium volume, a is the fcc lattice constant, n_0 is the mean density of electrons in $10^{-2} el/a_B^{-3}$; $N(0)$ is the density of states at the Fermi level in states/(spin Ry atom), l_{TF} is the Thomas-Fermi screening length, E_F is the Fermi energy (the occupied bandwidth), and k_F is the Fermi momentum.

[a] Here a is the experimental lattice constant at 35 GPa; our theoretical pressure is 30 GPa.

phonon spectrum is hardened), in fcc Li the Fermi surface evolves from a typical s-like sphere into a multiply-connected (Cu-like) shape, with necks extending through the L points, reflecting the increase in p character. In the phonon spectrum, structural instabilities appear around 35 GPa along the $\Gamma - K$ line, due to the strong e-ph coupling of some selected phonon modes, whose wave vector $\mathbf{q}$ connects the necks on the Fermi surface [55, 56]. We therefore expect that also the enatom of the two systems will display different behaviors under pressure.

5.2 Computational Details

The enatom could be computed by evaluating the linear response of a system to the displacement of a single atom, that is, the dielectric response $\epsilon(\mathbf{r}, \mathbf{r}') \leftrightarrow \epsilon(\mathbf{q} + \mathbf{G}, \mathbf{q} + \mathbf{G}')$. While this may be a viable approach, it is demanding and tedious and we use another method that requires only minor additional codes. Our approach is to let the computer do the linear response for us, by choosing a supercell, displacing one atom (taken to be at the origin), and obtaining the linear changes $\nabla_j n(\mathbf{r})$ and $\nabla_j v(\mathbf{r})$ by finite differences. Using cubic symmetry, displacement in a single direction is sufficient to obtain the full vector changes. The enatom components (rigid and deformation) are then obtained using the Helmholtz decomposition of a vector field into its irrotational and divergenceless parts [217].

We obtained the enatom for fcc Li and fcc Al at atomic volumes corresponding

5.3. RESULTS AND DISCUSSION

to 0, 35, and 50 GPa pressure. We used a cubic supercell of lattice constant $\mathcal{A}=3a$ (lattice constants are listed in Table 5.1), which contained $4\times3^3=108$ atoms. The enatom is represented as a Fourier series in the supercell. For the jellium calculations we set a single Li or Al atom into cubic supercells whose lattice constants correspond to the P=35 GPa cases. The mean electronic density was made equal to the related crystalline systems and a homogeneous positive jellium background provided charge neutrality.

For the self consistent density functional calculations we employed the PWSCF code [231] and Troullier-Martins [105] norm conserving LDA pseudopotentials and a plane wave cutoff energy of 20 Ry for both Li and Al. For the **k**-space integration in the primitive fcc unit cell we used a $(18)^3$ Monkhorst-Pack grid [232], with a cold–smearing parameter of 0.04 Ry [233]. With these parameters, we obtained a convergence of 0.2 mRy in the total energy and of 0.2 GPa for the pressure at 35 GPa for both systems. For the large cubic supercell, we used a 2^3 Monkhorst-Pack mesh, yielding four points in the irreducible Brillouin zone, and the same cold-smearing parameter of 0.04 Ry. With this choice, the total energy (pressure) calculated in the supercell equals that of the original fcc lattice within 0.1 mRy and 0.1 GPa, respectively. The pressure was calculated from a Birch-Murnagham fit of the LDA E versus V curve.[5]

In Table 5.1 we summarize the most relevant properties of Al and Li as a function of pressure. Since the independent variable in our calculations is the volume of the unit cell, we also include a column showing the relative volume change. We notice that the lattice constant of Li decreases very rapidly with pressure, as signaled also by the very small bulk modulus. At $P = 30$ GPa, the unit cell volume of Li is already one half of its $P = 0$ value. For comparison, the volume of Al at 50 GPa is 73% of its zero pressure value.

5.3 Results and Discussion

As a test of our numerical approach we checked that the sum rules of Eqs. (5.2) and (5.3) were almost perfectly fulfilled. For the jellium enatom the deformation part should be exactly zero, as discussed in Sec. 5.1.1. The deformation is not identically zero for our jellium enatom due to supercell effects. These effects are however very minor, *viz.* the maximum of the jellium deformation density is 500 times smaller

---

[5]To test the applicability of the pseudopotential method to high pressures, we calculated the energy versus volume relation for the two systems and fitted it to a Birch-Murnagham equation, to extract the equilibrium volume (V_0) and the bulk modulus at zero pressure (B_0) and its derivative at zero pressure (B'_0). For Lithium we obtained : $V_0 = 127.16$ $(a_B)^3$, $B_0 = 14.9$ GPa, $B'_0 = 3.33$. For Al we obtained: $V_0 = 105.3$ $(a_B)^3$, $B_0 = 82.7$ GPa, $B'_0 = 4.2$, in reasonable agreement with the experimental values ($V_0 = 111.2$, $B_0 = 72.7$ GPa, $B'_0 = 4.3$) [234]

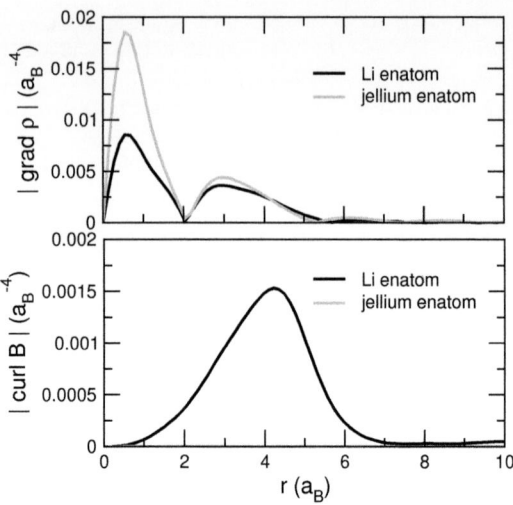

Figure 5.1: Comparison of the magnitude of the rigid (top) and deformation parts (bottom) in the first oder change of the density (Eq. (5.1)) of Li at 35 GPa and a jellium model. The plot is taken along a [110] direction through the atom. The magnitude of $|\nabla \times \mathbf{B}|$ in jellium is three orders of magnitude smaller than in Li, and therefore invisible on the scale of the plot.

than the maximum of the Li crystal deformation density.

5.3.1 Rigid and Deformation Part

In Fig. 5.1 we show the magnitudes of the vector fields $\nabla \rho$ and $\nabla \times \mathbf{B}$ along a [110] direction. The rigid parts for fcc Li and the Li-in-jellium model have the same magnitude and overall shape, while the deformation parts are very different. For Li $|\nabla \times \mathbf{B}|$ is approximately one order of magnitude smaller than $|\nabla \rho|$. This behavior reflects a general trend expected in simple metals: the deformation part is much smaller than the rigid, nevertheless it is quite insightful, as we demonstrate below.

To quantify the strength of the fields we are considering in a more precise way, we define the magnitude $\mathcal{M}[A]$ of a scalar or vector field $A(\mathbf{r})$ as the root mean square

$$\mathcal{M}[A] = \sqrt{\frac{1}{\Omega} \int_\Omega d^3\mathbf{r}\, [A(\mathbf{r})]^2}, \qquad (5.9)$$

5.3. RESULTS AND DISCUSSION

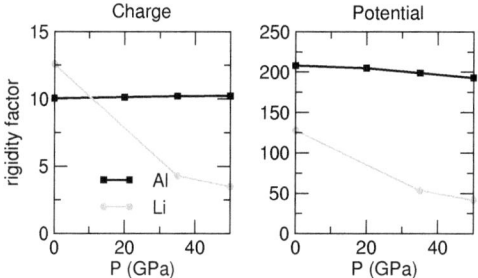

Figure 5.2: Evolution of the rigidity factor $\mathcal{R}$ (Eq. (5.10)) of the enatom for Li and Al as a function of pressure. The difference is the decrease in rigidity with pressure in Li, which is magnified somewhat by Li's larger compressibility.

where Ω is the volume of the supercell. The relative importance of $\nabla \rho$ and $\nabla \times \boldsymbol{B}$ in Eq. (5.1) can be quantified by defining a *rigidity factor* as

$$\mathcal{R} = \frac{\mathcal{M}[\nabla \rho]}{\mathcal{M}[\nabla \times \boldsymbol{B}]}. \qquad (5.10)$$

This ratio is one measure of how rigidly the enatom density (or potential, defined analogously) follows the nucleus. For the perfect jellium enatom $\mathcal{M}[\nabla \times \boldsymbol{B}] = 0$ and the rigidity factor would diverge. But because of the already mentioned supercell effects, we obtain $\mathcal{R} \sim 3000$ (1400) for the density and $\mathcal{R} \sim 2500$ (1500) for the potential of jellium Li (Al), which demonstrates again that the supercell effects are indeed small.

In the actual compounds, for the charge, $\mathcal{R}$ has similar values at zero pressure but decreases by a factor of 3 between 0 and 50 GPa in Li, whereas it is unchanging in Al, as shown in Fig. 5.2. For the potential, $\mathcal{R}$ is almost a factor of 2 smaller in Li than in Al at zero pressure, and again decreases with pressure while that for Al remains constant. This very different behavior in two simple metals is further confirmation that Li is increasing in covalency with pressure, while Al is not. The large values of $\mathcal{R}$ for the potential reflects the fact that the change in potential (e.g. due to phonons) is dominated by a rigid part, which provides justification for a rigid screened ion or the rigid muffin tin potential approximation [235].

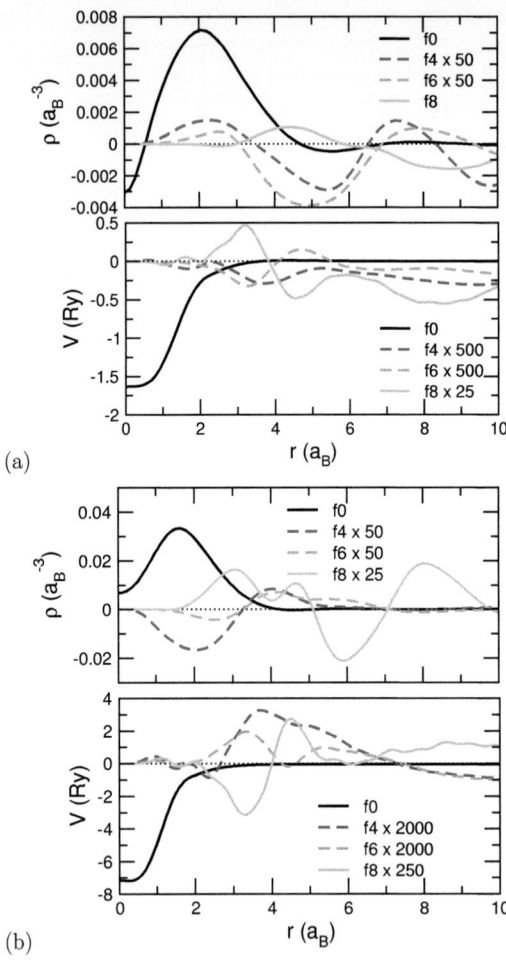

Figure 5.3: A kubic harmonic decomposition of the rigid parts ρ and V of the enatoms of (a) Li and (b) Al at 35 GPa. The radial expansion coefficients f_L are defined in Eq. (5.11).

5.3. RESULTS AND DISCUSSION

5.3.2 Rigid Part

Decomposition in Lattice Harmonics

To define the degree of sphericity of the rigid density ρ and potential V these scalar functions can be expanded in lattice harmonics of full cubic symmetry, identified with angular variation $L=0, 4, 6, 8...$:

$$\rho(\boldsymbol{r}) = \sum_L f_L(r) K_L(\hat{\boldsymbol{r}}) \qquad (5.11)$$

where K_L is the kubic harmonic [236] built from spherical harmonics of angular momentum L (see Ref. [237]), f_L are the radial expansion coefficients, and $r = |\boldsymbol{r}|$. In our case the functions K_L are normalized according to $\int (K_L)^2 d\Omega = 4\pi$, which ensures that the first radial expansion term is equivalent to the spherical average, i.e., $f_0(r) = 1/(4\pi) \int \rho(\boldsymbol{r}) d\Omega$.

In Fig. 5.3 we display the $L > 0$ radial expansion coefficients at 35 GPa, relative to the (obviously much larger) spherical part. The ideal jellium enatom is perfectly spherical and thus contains f_0 only (see Sec. 5.1.1). The decomposition of our supercell jellium model reveals small $L > 0$ terms, due to small supercell effects. The magnitudes of f_L for Li and Al are comparable to the supercell effects in the jellium enatoms, as the supercell boundary is approached. Therefore we conclude that the lattice harmonic coefficients f_4, f_6, f_8 are meaningful only within a radius of $\sim \mathcal{A}/3$.

While the general characteristics and relative signs of the $L > 0$ terms are quite different in Li and Al, in most cases their maximum is only $\sim 1\%$ or less of the maximum of f_0, for both density and potential. The only exception is the $L = 8$ lattice harmonic in the density of Li, which is surprisingly large around the nearest neighbor distance 4.5 a_B. This anisotropy reflects the 'cooperative' influence of neighboring atoms in determining the enatom character. The relative size of the $L = 8$ peak grows from 8% to 16% of the maximum of f_0 from 0 to 50 GPa. With increasing covalency in Li under pressure not only f_8 but all non-spherical contributions to the rigid density increase; this effect cannot be seen in Al.

For the enatom potentials, the non-spherical terms are small enough in fcc Li and Al at all volumes studied that the enatom potentials can be considered effectively spherical, as is the common assumption in simple metals.[99]

Effects of Screening

Figures 5.4 and 5.5 show the rigid parts of density and potential and their pressure evolution. From linear screening theory we know the spherical part of the induced change in charge density $\Delta n(r)$ for a simple metal will have long-range (but rapidly decaying) Friedel oscillations. Our approach reproduces the long-range oscillations

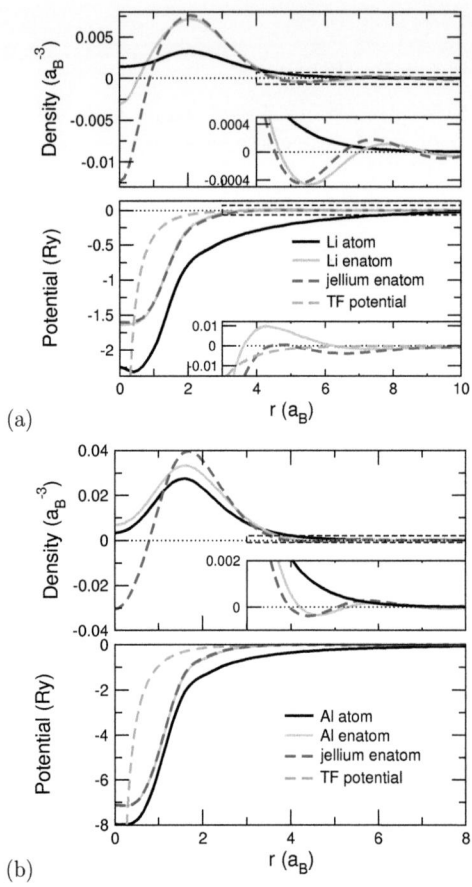

Figure 5.4: Radial plots of the rigid parts of density and (local, $l = 2$) potential for (a) Li and (b) Al, both at 35 GPa. We show the isolated atom, the enatom, the enatom of a jellium model, and, only in the lower panels, the Thomas-Fermi potential. The plots show the spherical parts f_0 of these quantities. On the radial direction the eight nearest neighbors are located at (a) 4.53, (b) 4.87 a_B, the six next nearest neighbors are at (a) 6.41, (b) 6.89 a_B and the edge of the supercell is at (a) 13.59, (b) 14.62 a_B, illustrating the very small overlap from neighboring supercells. In the inset we show a blow-up of the tail region, where the Friedel oscillations of the charge are clearly visible (see also inset in Fig. 5.5).

5.3. RESULTS AND DISCUSSION

according to

$$\Delta n(r) \sim \cos(2k_F r)/r^3. \tag{5.12}$$

with reasonable agreement (see insets in Figs. 5.4 and 5.5); they should be important primarily for describing long-range force constants. The corresponding oscillations of the enatom potential are not expected to be as important, and we confirm this expectation, as the oscillations visible in the inset in the lower panel of Fig. 5.4(a) are indeed very small.

In the lower panels of Fig. 5.4 we also see that the screening in the solid causes the enatom potentials to be significantly more short ranged than the atomic potentials. Furthermore, the enatom potential is less attractive by 0.7 to 0.9 Ry in these atoms. The gradients, which determine electron-phonon matrix elements, do not seem to differ greatly. Note that the pseudopotential we have used is non-local, and in the plots only the local ($\ell = 2$) component is shown. The total potential will include the $\ell = 0, 1$ nonlocal parts, which are non-vanishing only within the core radius ($r_{\text{core}} \simeq 2$ a_B) and move rigidly.

To understand this screening better we compare the rigid enatom potential V with the Thomas-Fermi potential

$$V_{\text{TF}}(r) = -\frac{Q^{\text{val}}}{r} \cdot e^{-r/l_{\text{TF}}}, \tag{5.13}$$

where $Q^{\text{val}} = eZ^{\text{val}}$ is the total charge of the valence electrons and l_{TF} is the Thomas-Fermi screening length calculated from the mean valence density[6] (given in Table 5.1). For both systems and all pressures we find good agreement for the long range behavior, confirming that the system is still dominated by homogeneous electron-gas screening. The electronic density of Al is higher than the one of Li and therefore the screening is stronger. As a consequence the effective potential in Al is more localized than the one of Li. The agreement with linear screening is better for Al than for Li, further supporting the deviation of Li from the homogeneous electron density picture.

Rigid Density and Rigid Potential

The spherical average $f_0(r)$ of the rigid enatom density ρ contains a charge equal to the valence, which is compared to a valence density for the isolated atom in Fig. 5.4, and also with the corresponding enatom in jellium of the appropriate density. Note first that, while in a pseudopotential calculation the density (potential) inside the

---

[6]Thomas-Fermi screening length: $l_{\text{TF}} = 1/2(\pi/3n_0)^{1/6}$, n_0 is the mean electronic valence density in atomic units (see Table 5.1).

core radius does not have much physical meaning, changes within the core radius will still be useful probes of the enatom character.

The enatom density and potential are both more localized than in the isolated atom. This difference can be attributed to two effects.

(i) The density in the tail region is screened in the solid, making the effective potential more short ranged and causing charge to move inward. As a result the peak value around 2 a_B increases.

(ii) There is also charge that moves outwards from the core region, causing the peak value to increase further but also to move outwards.

The tail is very similar in jellium and in the solid (a consequence of similar Thomas-Fermi screening), but in the core region and around the maximum the densities are different. The jellium enatom density becomes negative near the nucleus, with the amount of negative density in the core region being about 1% of the valence, for both the Li and Al jellium models. Thus this region does not contain a significant amount of negative density, but it still causes some charge to move away from the core. As a result the peak value of the jellium enatom is slightly higher and further out than the one of the crystal enatom. (Note: in actual supercell calculations there is never any negative density, as the negative dip is compensated by tails of neighboring atoms.)

Evolution under Pressure

The pressure evolution of the rigid enatom density and potential are compared on an absolute length scale in Fig. 5.5. Here we can identify aspects of the same two effects as described in the preceding section. Under pressure the enhanced electronic screening makes the effective potentials more short ranged, and results in the screened charge moving inward (effect (i)). For Al the increasing pressure causes first a decrease of the extent of rigid density around 4 a_B (see inset) and second, an increase of the peak value around 2 a_B of about 12% from 0 to 50 GPa. For Al the potential change with pressure is negligible.

Effect (i) (screening) has a stronger influence in Li than in Al because, first, the density is lower so screening is less, and second, the relative volume change is larger. The peak value increases by more than 1/3 from 0 to 50 GPa.

But here we also find effect (ii) which causes a small amount of charge to move away from the core and leads to an outward shift of the peak position in the rigid density from 1.9 a_B (P=0) to 2.1 a_B (50 GPa). The shift of charge is indicated more clearly in Fig. 5.6, where the integrated charge is shown for Li.

The pressure evolution of the enatom potential is characterized by a decrease in attraction at small r, from -1.78 Ry (P=0) to -1.60 Ry (50 GPa) in Li. This decrease is fairly uniform over the region out to 3.2-3.5 a_B beyond which it becomes negligible. In Al the decrease in attraction between P=0 and P=50 GPa is only 2%. For 35 and

5.3. RESULTS AND DISCUSSION

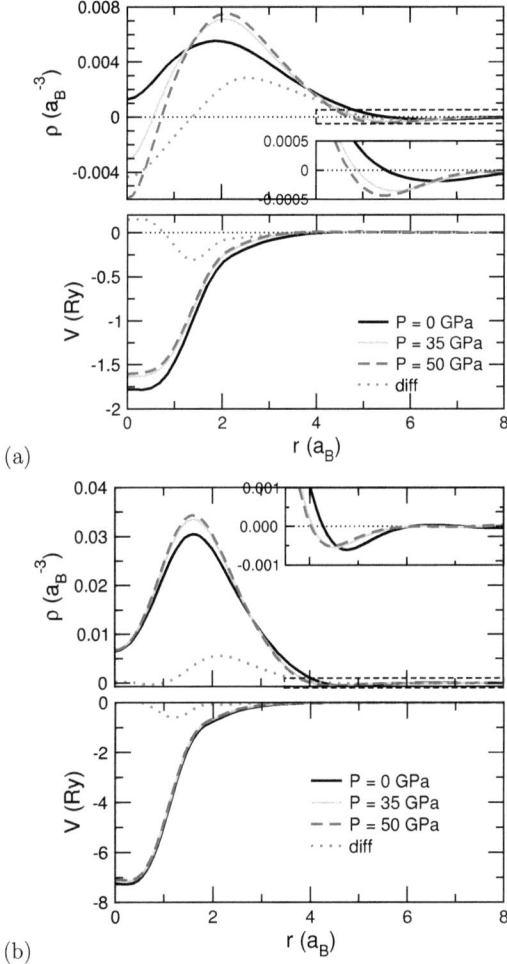

(a)

(b)

Figure 5.5: The pressure evolution of the rigid part of the enatom in (a) Li and (b) Al, plotted along the [110] direction. The dotted line represents the difference between the quantities calculated at P=35 GPa and those at P=0. Under pressure, the enatom density increases at its peak (at 1.5-2 a_B), and the local ($l = 2$) potential becomes less attractive due to the increased screening.

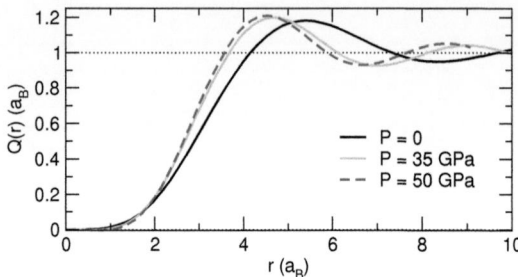

Figure 5.6: Integrated spherical density $Q(r) = 4\pi \int_0^r r'^2 f_0(r') dr'$ for Li at 0, 35, and 50 GPa. The density is shown in Fig. 5.5. For $r > 2\ a_B$ the shift in density inward with pressure is evident, although there is little difference between 35 and 50 GPa. For $r < 2\ a_B$ a small amount of charge is also shifted outward.

50 GPa the rigid density of the Li enatom is negative inside $1\ a_B$, but the amount of negative density contained within that region is less than 1% compared to the total valence charge, and does not even show up in the plot of the integrated charge in Fig. 5.6.

5.3.3 Deformation part

In general the vector fields $\boldsymbol{B}$ and $\boldsymbol{W}$ (Eqs. (5.4) and (5.5)) that describe the deformation parts of density and potential have similar morphologies that reflect the symmetry of the lattice. $\boldsymbol{B}$ or $\boldsymbol{W}$ form symmetry related "donut" swirls centered at different distances along lines connecting the central atom and first (1nn) and second (2nn) nearest neighbors, i.e., the crystal axes (see Fig. 5.7(a), where only swirls around the 2nn are visible). The swirls associated with the 1nn and 2nn have opposite rotational directions. The derived fields $\nabla \times \boldsymbol{B}$ or $\nabla \times \boldsymbol{W}$ are large at the centers of the swirls of $\boldsymbol{B}$ or $\boldsymbol{W}$ (see Fig. 5.7(c)), and they are primarily directed radially.

It is informative also to view these fields in planes as done in Fig. 5.7(b) or Fig. 5.7(d), where the precise position and spatial extent of their features can be judged. It can be seen, for example, that the donuts pictured in Fig. 5.7(a) are nearly centered on the 2nn Li sites and that $\boldsymbol{W}$ is oriented perpendicular to the plane, pointing either towards the viewer (vectors visible) or in the opposite direction (vectors not visible). As only the fields $\nabla \times \boldsymbol{B}$ or $\nabla \times \boldsymbol{W}$ are involved in the calculation of physical properties (see Eqs. (5.1), (5.4), and (5.5)), we will focus our attention on them. A comparison of the deformation parts of density and potential for two

5.3. RESULTS AND DISCUSSION

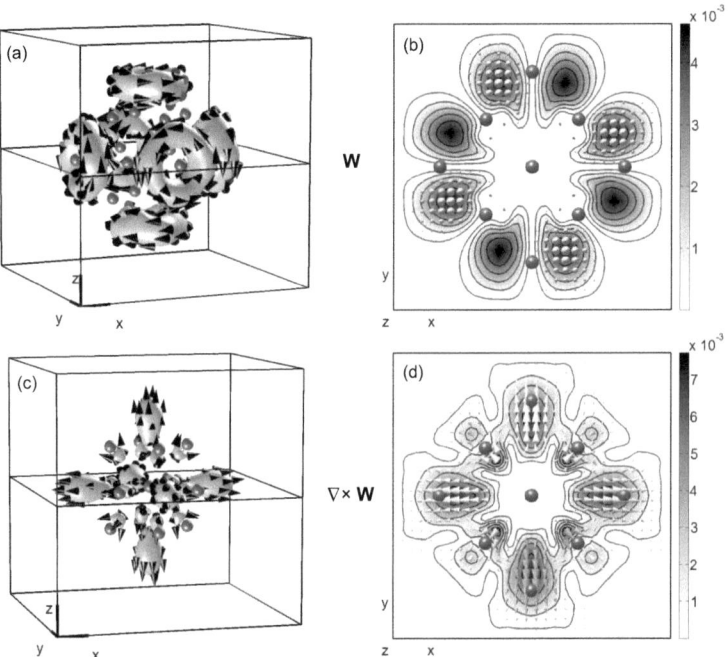

Figure 5.7: The deformation part of the *local potential* of Li at P = 35 GPa. (Left-hand panels) Three dimension isocontour graphs of the magnitude, with arrows indicating the direction; (right-hand panels) contour plots in the (001) plane. (a) and (b) show the vector field $\boldsymbol{W}$, (c) and (d) $\nabla \times \boldsymbol{W}$. The dark green (dark gray) balls represent the position of the central atom and the nearest and next nearest neighbors within the supercell, which is displayed as black boundary box. The orange (light gray) isocontours in (a) and (c) indicate $|\boldsymbol{W}| = 3.5 \times 10^{-3}$ and $|\nabla \times \boldsymbol{W}| = 4.8 \times 10^{-3}$, respectively. The black arrows are field vectors that are located on the isocontours. (b) and (d) indicate the magnitude of the vector fields within xy-planes that are indicated as black-lined squares in the 3D graphs. Superimposed is a mesh of yellow (light gray) field vectors which are located within the plane.

different pressures, for both Li and Al, is given in Figs. 5.8 and 5.9.

Deformation Part of the Density

In Li at P=0 the maxima of the charge deformation $\nabla \times \boldsymbol{B}$, shown in Fig. 5.8, are strongly localized around the nearest neighbors. Under pressure these maxima are pulled inward. The direction of the field determines the sign of the charge deformation. For Li in Fig. 5.8 $\delta \boldsymbol{R} \cdot \nabla \times \boldsymbol{B}$ (see Eq. (5.4)) is positive for $\delta \boldsymbol{R} \parallel [100]$, so there is a 'charge transfer' from behind the displaced atom, to in front of it. Such charge distribution reflects a displacement-induced dipolar moment described by the deformation (and which will be screened locally in a metal). At ±45° (at the 1nn sites, in fact) there is a depletion of charge, with a corresponding increase at ±135° (on the 1nn behind the displaced atom). In Al the pattern is similar, but the sign is reversed and the maxima are nearer the nucleus. These differences will affect their dynamical properties differently; this influence may be significant in Li and is probably not in Al as it remains more free-electron-like.

Under pressure, the magnitude of $\mathcal{M}[\nabla \times \boldsymbol{B}]$ in Li increases quite significantly, being 2.8×10^{-5} at P=0 and increasing by an order of magnitude at 50 GPa. This change, consistent with increased covalency, is the cause of the large drop of the rigidity factor in Fig. 5.2. The pressure evolution in Al is marginal: $\mathcal{M}[\nabla \times \boldsymbol{B}]$ is 1.9×10^{-4} at P=0 and increase by only 30% at 50 GPa.

Deformation Part of the Potential

The potential deformation in Li undergoes a surprisingly large pressure evolution, reflected in the shape, magnitude $\mathcal{M}$, and extent of $\nabla \times \boldsymbol{W}$. $\mathcal{M}[\nabla \times \boldsymbol{W}]$ is 4.4×10^{-4} at P=0 and increases by over a factor of 4 by 50 GPa. The contour plot of Fig. 5.9 shows the change from P = 0 to 35 GPa. Starting with a small deformation located on the 1nn, maxima in the deformation grow in substantial regions including the 2nn.

For Al, $\nabla \times \boldsymbol{W}$ has its maxima along the cubic axes, and much closer to the nucleus. As for the charge deformation, $\nabla \times \boldsymbol{W}$ has the opposite sign compared to Li, and its change with pressure is minor.

Given the simple shape of the deformation term $\nabla \times \boldsymbol{W}$, it is easy to understand its effect on the total change in the potential (Eq. (5.5)). $\delta \boldsymbol{R} \cdot \nabla \times \boldsymbol{W}$ gives an additional dipolar-type contribution, adding to the main change of potential $\nabla_j v$ which has a dipolar form arising from displacement of the (nearly spherical) rigid potential.

The pressure evolution of the rigidity factor for the potential in Li (see Fig. 5.2) shows that at 50 GPa the deformation part contributes about 2% to the total change in potential $\nabla_j v$ (for Al this contribution is negligible). For materials with lower

5.3. RESULTS AND DISCUSSION

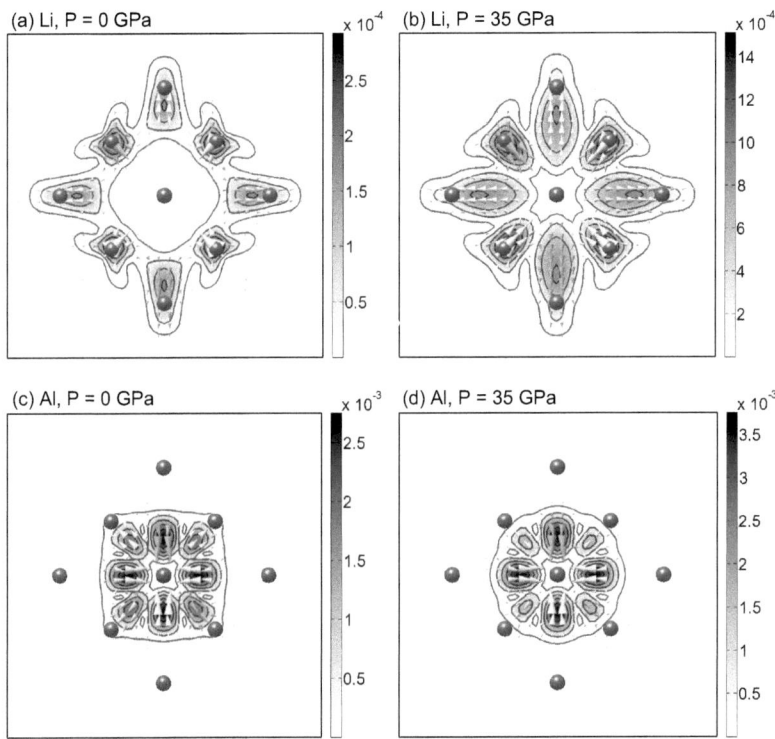

Figure 5.8: 2D graphs (see Fig. 5.7) of the deformation part of the *charge density* $\nabla \times \boldsymbol{B}$ for Li and Al for 0 and 35 GPa. Li undergoes a significant pressure evolution arising in the shape and the magnitude of $\nabla \times \boldsymbol{B}$. Al in turn changes only slightly. For the meaning of the symbols see Fig. 5.7.

rigidity the deformation part might give substantial contributions to $\nabla_j v$, large enough to affect its scattering properties or the strength of electron-phonon coupling.

Additional to the figures shown in this chapter we provide several color graphs, showing examples of enatom quantities for Li in 3D and 2D views, in appendix A.

5.4 Summary and Outlook

In this chapter we have provided a numerical linear response approach, and the first explicit examples, of the *enatom* (the generalized pseudoatom introduced by Ball [50]) density and potential for Li and Al, at pressures of 0, 35, and 50 GPa. The enatom consists of a rigid and a deformation density (and potential). The rigid part defines a unique decomposition of the equilibrium density (potential) into atomic–like but overlapping contributions that, to first order, move rigidly with the nuclear position. The deformation density (potential) describes how this charge (potential) deforms, and it can be viewed as a backflow, or (depending on its shape) as a mechanism that transfers charge from one side of the displaced atom to the other. The enatom quantities were obtained from supercell finite–difference calculations, demonstrating that this approach provides a feasible numerical treatment.

A *rigidity factor* $\mathcal{R}$ was introduced to quantify the relative importance of the rigid and deformation parts of the enatom. It characterizes how rigidly the charge (or potential) moves upon displacement. The rigidity factor is expected to be small for covalent materials whose bonding is strongly direction-dependent, and large for metals that lack such a strong bonding [50]. It has been shown recently that Li becomes more covalent under pressure [56, 57] and the different components of the Li enatom have confirmed this trend. Aluminum, on the other hand, remains quite free-electron like up to 50 GPa. Both behaviors are clearly reflected in the pressure evolution of the rigidity factor of both the density and the potential: $\mathcal{R}$ decreases by a factor of $3-4$ in Li but stays almost constant in Al. Moreover, the rigidity of the potential is approximately one order of magnitude bigger than for the density. This rigidity supports the picture of a rigid potential shift with displacement in both Li and Al. Therefore, the rigidity factor $\mathcal{R}$ may become a useful tool for quantifying a "generalized covalency" of a system, even in the case of metals.

By kubic harmonic decomposition of the rigid enatom, we have shown that in Li and Al the potentials are effectively spherical, supporting spherical approximations in rigid–atom models of electron–phonon coupling. Non–spherical contributions in the rigid density become larger as Li becomes more covalent under pressure. For aluminum, changes are much smaller.

The basic features of the spherical part of the rigid density and potential can be understood by means of linear screening theory. First, the localization of the rigid

5.4. SUMMARY AND OUTLOOK

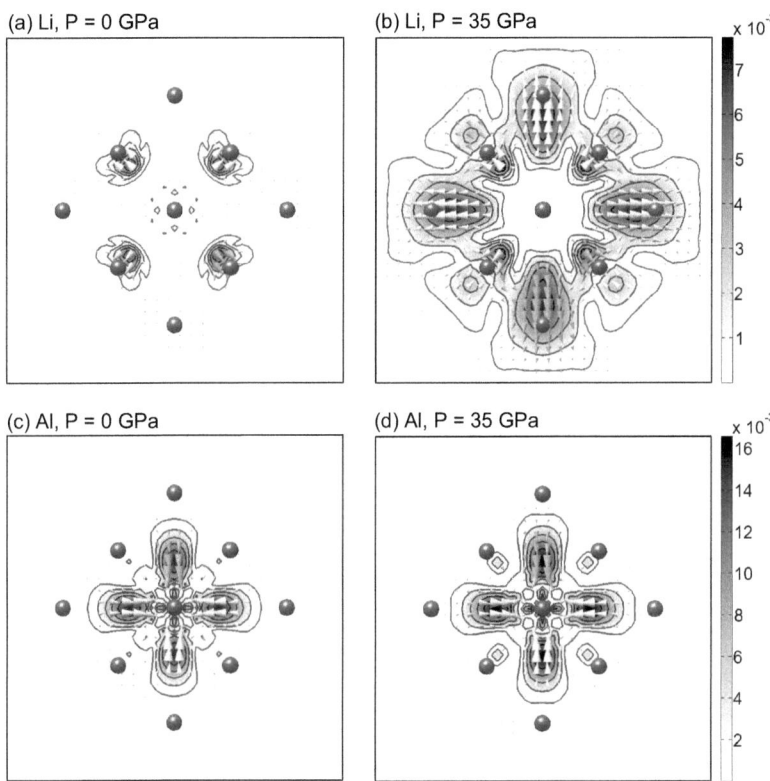

Figure 5.9: 2D graphs (see Fig. 5.7) of the deformation part of the (local) *potential* $\nabla \times \boldsymbol{W}$ for Li and Al for 0 and 35 GPa. Plots (a),(b) and (c),(d) share the same color bar, respectively. Lithium undergoes a significant change with pressure, while aluminum remains almost unchanged. For the meaning of the symbols see Fig. 5.7.

enatom potential is a result of free electron-like (Thomas-Fermi) screening, showing that the mean radius of Al is smaller than the one of Li. Second, the tails of the rigid densities exhibit rapidly decaying Friedel oscillations.

Another finding is that the rigid enatom density is more localized than the density of an isolated atom. This is a result of two effects. (i) The density in the tail region is screened in the solid, making the effective potential more short ranged and causing charge to move inward. (ii) There is also charge that moves outwards from the core region, which is an effect of the potential being less deep than in the free atom. This second effect is small compared to the first one. Both effects also cause an increase of localization of the rigid enatom density when the pressure is increased.

The pressure evolution of the deformation part of the enatom is quite significant in Li, but small in Al. The basic morphological features of the deformation parts are the same in Al and in Li as well as in the density and in the potential, but their position, sign, and relative magnitude are different. Such a behavior confirms the expectation that the lattice symmetry determines the character of the deformation of the enatom, at least in nearly free electron metals.

We conclude that the numerical procedures presented above provide a viable route to determine enatom quantities. As the enatom is intrinsically a dynamically determined quantity (involving linear response), it allows for the calculation of dynamical properties such as phonon dispersions and electron–phonon matrix elements. The method thus provides a real–space picture of solids and their dynamical properties, which could contribute to a deeper understanding of the physical properties of materials. We foresee the important application of the method in the electron-phonon problem and the study of strongly coupled elemental metals and compounds. For further applications, our method ought to be generalized to systems with many atoms per unit cell and with non–cubic symmetry. As our understanding of the enatom is still at an early stage, studies of non–metals, i.e., covalent or ionic materials, should be carried out in the future.

Chapter 6
Summary and Outlook

This thesis is dedicated to the theoretical study of *sp* materials, in particular boron, lithium, and aluminum. We employ density functional theory (DFT) and density functional perturbation theory (linear response) within the framework of the local density approximation (LDA) or the generalized gradient approximation (GGA). Those approximations are well suited for the description of *sp* materials and allow us to calculate a variety of material properties. All theoretical methods are described in chapter 2.

The basic motivation of this thesis is to improve our understanding of the properties of materials that are based on elemental boron. We are interested in elemental boron because it has a fascinating chemical and structural complexity, it is little studied and many fundamental properties are still unknown. The main body of this thesis is presented in chapter 3, where we study different structural, electronic, mechanical, thermodynamic, and vibrational properties, the chemical bonding, and electron–phonon interactions of multiple nanomaterials and bulk phases of boron.

The starting point of our studies is the Boustani Aufbau principle (see introduction or Sec. 3.2.3). It is a very general scheme that predicts the existence of quasiplanar (sheets), tubular (nanotubes), convex and spherical (fullerenes) boron clusters. The Aufbau principle is based on extensive numerical studies of small boron clusters, employing quantum chemical methods like DFT. After these predictions, it was experimentally found that small boron clusters indeed from sheet–like, quasiplanar structures [19, 20] and also boron nanotubes (BNTs) were synthesized [21]. These experiments confirm the validity of the Boustani Aufbau principle and also show the predictive power of DFT. At the same time they open the door to a new field of science based on boron nanostructures. However, the Aufbau principle is a very general approach and the corresponding experimental studies are not very detailed yet. Therefore questions about the precise atomic structure of boron nanotubes and boron sheets remain open, and further theories describing their properties are needed.

We provide such a theory in chapter 3. It is based on the fact that much of the physics of carbon fullerenes and carbon nanotubes can be understood in terms of graphene, a broad carbon sheet. Therefore graphene can be considered as the precursor of carbon fullerenes and carbon nanotubes. The existence of BNTs raises the question if such a broad sheet also exists for boron. This sheet would be the limiting case of the experimentally observed quasiplanar clusters for an infinite number of atoms, and the precursor of BNTs. The experimental verification of the Boustani Aufbau principle has shown that for boron materials DFT is a reliable tool, being able to predict new materials. We therefore use DFT to determine the structure and the properties of that broad boron sheet (BS), that is considered here for the first time. We then apply our findings to the related boron nanotubes in order to predict their basic properties. Before, BNTs and BSs were mainly studied in the context of the Boustani Aufbau principle as finite sized clusters [23, 24, 177, 110]. Our findings in chapter 3 unify the results of these former studies into a generalized theory of BSs and BNTs.

To determine the structure of the BS, we examine a number of different structure models. By DFT simulations we find that a sheet with a triangular boron lattice and a simple up–and–down puckering is the most stable one and is very likely to be the precursor of BNTs (see Fig. 3.12 on page 71). At the moment in which we carried out these studies the same BS structure was found by two other groups [173, 174]. This independently confirms the present structure model and underlines the scientific interest in boron based materials. This metallic BS is the cental structure of chapter 3. The sheet is held together by homogeneous multi–center bonds and by linear sp hybridized σ bonds exclusively lying along the armchair direction of the sheet. The anisotropic bond properties of the sheet lead to different elastic moduli C_x and C_y for stretching the BS in the x and in the y direction. Furthermore, puckering of the BS can be understood as a key mechanism to stabilize the sp σ bonds.

We then study *ideal* BNTs, that are constructed from this BS by a "cut–and–paste" procedure. We predict the existence of helical currents in ideal chiral BNTs, which means that chiral BNTs are nanocoils. Furthermore, we show that all ideal BNTs are metallic, irrespective of their structure. BNTs could therefore be perfect nanowires. However, we find that ideal BNTs do not represent the ground state of BNTs, and we identify structures of lower symmetry (we called them *real* BNTs), which are higher in cohesive energy. We show that real BNTs have strain energies that depend on the nanotube's radius as well as on the chiral angle. This is a unique property among all nanotubular materials reported so far, and the implications of this finding were further explored in chapter 4.

We conclude that our findings in chapter 3 define a consistent picture of BSs and BNTs and unify former studies on these materials [23, 24, 177, 110] into a generalized theory. Independent of us other groups also studied BSs and BNTs [173, 174, 175, 238]. However, our study (published in Refs. [178, 239]) is by far the most extensive one in the field. While writing up this thesis the field of boron nanomaterials has evolved further. Szwacki et al. [33] proposed a model for a particularly stable spherical cluster (boron fullerene) and based on this model Tang et al. [196] and Yang et al. [197] revised our theory of BSs and BNTs. Overall the field has received considerable attention in the media as recent articles in scientific newspapers reveal [198, 199, 200, 201, 202, 203].

One of the main predictions of our theory of BNTs is that the strain energy of BNTs is a function of the nanotube's radius and chirality. In chapter 4 we examine this finding in detail and propose a new route to achieve control over the atomic structure of nanotubes during their synthesis. By analyzing the unfavorable case of carbon nanotubes, we show that our current inability to control their chirality is caused by isotropic in–plane mechanical properties of the related graphene sheets, leading to isoenergetic nanotubes with similar radius, but a whole range of different chiral angles. This "degeneracy" is lifted for nanotubes that are derived from a reference sheet with anisotropic in–plane mechanical properties. As demonstrated for the case of BNTs, this anisotropy will make the different chiral angles energetically separable. And this should be experimentally accessible during the synthesis process. Thus, in order to achieve a higher degree of structural control over nanotubular materials, we propose that one should systematically search for nanotubes which are related to sheets with anisotropic in–plane mechanical properties. The results of chapter 4 were published in Ref. [240].

In the first parts of chapter 3 we show that a broad boron sheet can be used to predict structural, electronic, and mechanical properties of boron nanotubes. Although it has not been observed in experiment yet, one might still wonder whether the BS has some further significance beyond its relation to boron nanotubes. Similar to carbon, where a stacking of many graphene layers generates the crystal structure of graphite, a stacking of BSs would constitute layered boron bulk materials. The pronounced polymorphism [35] and the unknown phase diagram [10] of elemental boron makes it quite probable that novel, so far undiscovered phases exist. If these structures are not stable at ambient conditions, they might be so at high pressures.

To find out whether such layered bulk phases exist, the three structures Immm, Fmmm, and α–Ga are examined in the remaining parts of chapter 3. The Immm phases is constructed as a stacking of the BS, discussed above, and is considered here for the first time. The Fmmm phase was proposed by Boustani et al. [177], who were the first to study layered bulk phases of boron. The phase α–Ga is boron in the α–gallium structure [44, 45, 46], and we find that it falls into the class of layered materials. In these three structures each boron atom is primarily coordinated within a quasiplanar layer, where the bonding is of three–center type, and it has at most one (two–center) σ bond between two layers. Furthermore, all these structures have metallic properties. Because the three phases have a similar structure and the same basic bonding pattern, they constitute a new *family* of (hypothetical) layered boron bulk materials. The existence of this family of boron structures is postulated here for the first time.

From a structural viewpoint the layered systems are very different from the common bulk phases, that consist of complex three–dimensional networks of B_{12} icosahedra. However, we find that the chemical bonding is quite similar in the icosahedral and layered phases. These similarities support the existence of layered boron phases, and allow us to formulate the following generalized picture of the chemical bonding in elemental boron solids: A three-center bonded triangular network of boron atoms forms basic units (icosahedra or quasiplanar layers) that are interconnected via σ bonds. Our generalization could be the starting point for a deeper understanding of chemical bonding in boron solids.

In the study of bulk structures, we note that elemental boron, a semiconductor at ambient conditions, transforms to a superconductor under pressure [9]. The biggest problem for explaining the superconductivity in boron is the general lack of knowledge about its high–pressure phases, and the corresponding crystal structures are unclear. Therefore, the problem of superconductivity merges with that of studying bulk structures under pressure. So far, three different theoretical approaches were used to determine possible high–pressure phases. One is based on studying the high–pressure behavior of the common icosahedral bulk structures [39, 40], another on randomly trying different naive phases such as fcc, bcc, etc. [41, 42, 43], and a third approach assumes that boron under pressure adopts similar structures than heavier group-III elements (Al, Ga, In) [44, 45, 46]. Up to now, however, the problem was not definitely solved. We approach the problem of high–pressure boron from a different side and study the layered bulk phases, which are derived from our insights about the chemical bonding in boron solids.

In order to judge the thermodynamic stability of the three layered systems we compare them with α–rhombohedral boron (R–12) and face–centered cubic (fcc) boron and calculate the $T = 0$ K phase diagram. R–12 is the simplest icosahedral phase and fcc is a simple closed packed structure. The phase diagram shows that between about 100 and 600 GPa α–Ga is the thermodynamically most favorable structure, and between about 150 and 300 GPa both α–Ga and Immm are favorable. The latter is exactly the pressure range where boron was experimentally found to be superconducting. These results indicate that layered bulk materials of boron could exist and that they are prominent candidates to explain the superconductivity in boron.

Therefore we study the electronic and phononic structure, as well as the electron–phonon coupling of the three layered phases in detail. The corresponding electronic band structures are fully three-dimensional, which shows that the phases are "layered" only in a geometrical sense. Boron in the α–Ga structure is dynamically stable at $P = 0$ and 210 GPa (all phonon frequencies are real). This finally is the strongest indication that layered bulk phases of boron can exist. At $P = 0$ GPa we predict that α–Ga is a $T_c = 2$ to 6 K conventional superconductor, and at 210 GPa strong electron–phonon coupling should lead to a measurable superconductivity. Boron in the Immm structure is dynamically unstable at $P = 0$ and 210 GPa (there are imaginary phonon frequencies). Nevertheless, we expect the phase to be stable at intermediate pressures or in a modulated superstructure at 0 GPa. At ambient conditions the system would be a 7 to 16 K conventional superconductor if we ignore the slight instability. The Fmmm structure is thermodynamically unfavorable and dynamically unstable. It can therefore be ruled out as a possible allotrope of elemental boron at ambient and high pressures.

In summary, we show that the three layered bulk phases exhibit chemical bonding patterns which are typical for boron solids, that α–Ga and Immm are thermodynamically favorable under pressure, and that α–Ga is dynamically stable at $P = 0$ and 210 GPa. We thus conclude that novel layered bulk phases of boron are likely to exist at elevated pressures or even at ambient conditions. Furthermore, there are strong indications that these layered phases are conventional superconductors. However, our present results are not able to unravel the origin of the experimentally reported high–pressure superconductivity in elemental boron.

To further improve our understanding of the layered boron bulk phases and to investigate their possible relation to the experimentally observed high–pressure superconductivity, future studies should stabilize the Immm phase by studying a superstructure or intermediate pressures, determine the pressure dependence of the electron–phonon coupling of the layered phases in more detail by considering more pressure points (atomic volumes), and put the yet qualitative results for the electron–phonon coupling in α–Ga at 210 GPa to a quantitative level. Furthermore, there are several indications that a common icosahedral structure of boron, just as α–rhombohedral boron, could also be responsible for the observed high-pressure superconductivity. Therefore, studying electron–phonon coupling in compressed icosahedral phases should be addressed. In general, the electronic structure and the chemical bonding of the elemental phases of boron ought to be analyzed in more detail and simple orbital–based descriptions should be developed.

For our studies of the layered bulk phases and the superconductivity in boron, linear response calculations (in the framework of density functional perturbation theory [47, 48, 49]) have proven to be an efficient way to determine dynamical properties of solids. However these methods are usually based on an abstract reciprocal space formulation and do not allow us to gain a direct understanding of the relations between structure, bonding, phonons and electron–phonon coupling. A method that provides a real–space picture of of a solid, its vibrational properties, and the electron–phonon interactions was proposed in the 1970s by M. A. Ball [50, 51], but never applied. It allows to describe condensed matter as a collection of generalized pseudoatoms, which we call *enatoms*.

In chapter 5 we present the first realization of this method. We develop a technique to construct the enatom, provide the first explicit examples of enatom quantities and analyze their properties. Rather than starting from boron, which is quite complex, we first apply the enatom method to the simple systems fcc lithium and fcc aluminum. These simple metals show different physical behaviors under pressure, which reflects the increasing covalency in Li and its absence in Al. The pressure evolution of the different components of the enatoms of Li and Al clearly reflect this trend. The results of chapter 5 were published in Ref. [241].

For further applications, our method should be generalized to systems with many atoms per unit cell and with non–cubic symmetry. Our understanding of the enatom is still at an early stage and studies of non–metals, i.e., covalent or ionic materials, must follow. In a further step dynamical properties such as phonon dispersions and electron–phonon matrix elements ought to be calculated. Finally, we would like to apply the enatom method to boron and study the relation between structure, bonding, and the dynamical properties in its various structures.

Appendix A
Enatom Quantities

The enatom method was introduced in chapter 5. This appendix shows examples of enatom quantities for lithium in 3D and 2D views. These are

- the rigid part of the enatom in lithium at P = 35 GPa,
- the deformation part of the enatom in lithium at P = 35 GPa,
- the pressure evolution of the enatom density in lithium,
- the pressure evolution of the enatom potential in lithium.

The *Rigid Part* of the Enatom in Lithium at P = 35 GPa

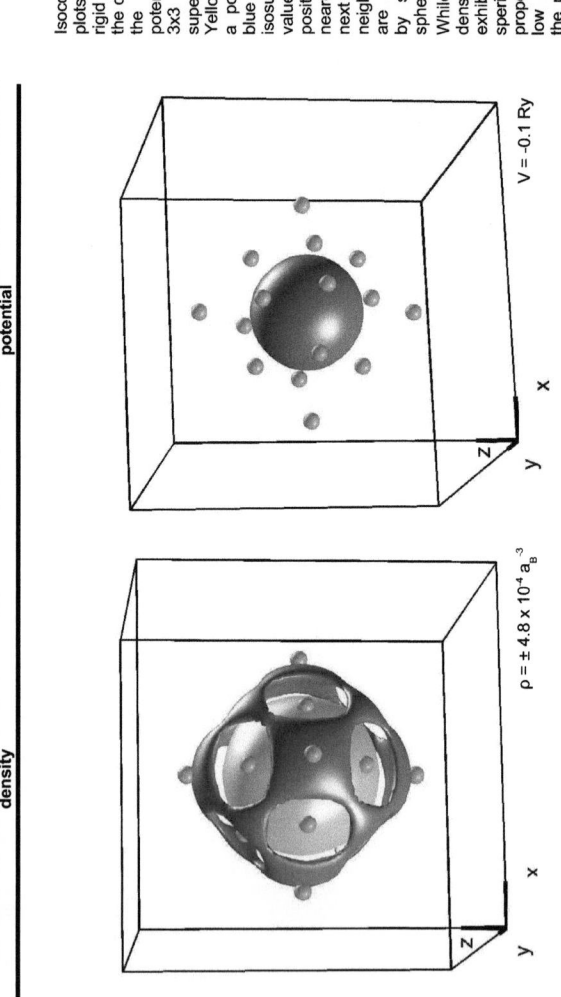

Isocontour plots of the rigid parts of the density and the local potential in the 3x3 cubic supercell. Yellow denotes a positive and blue a negative isosurface value. The positions of nearest and next nearest neigbor atoms are indicated by small pink spheres. While the density clearly exhibits non-sperical properties at low contours, the potential is much more nearly spherical.

The *Deformation Part* of the Enatom in Lithium at P = 35 GPa

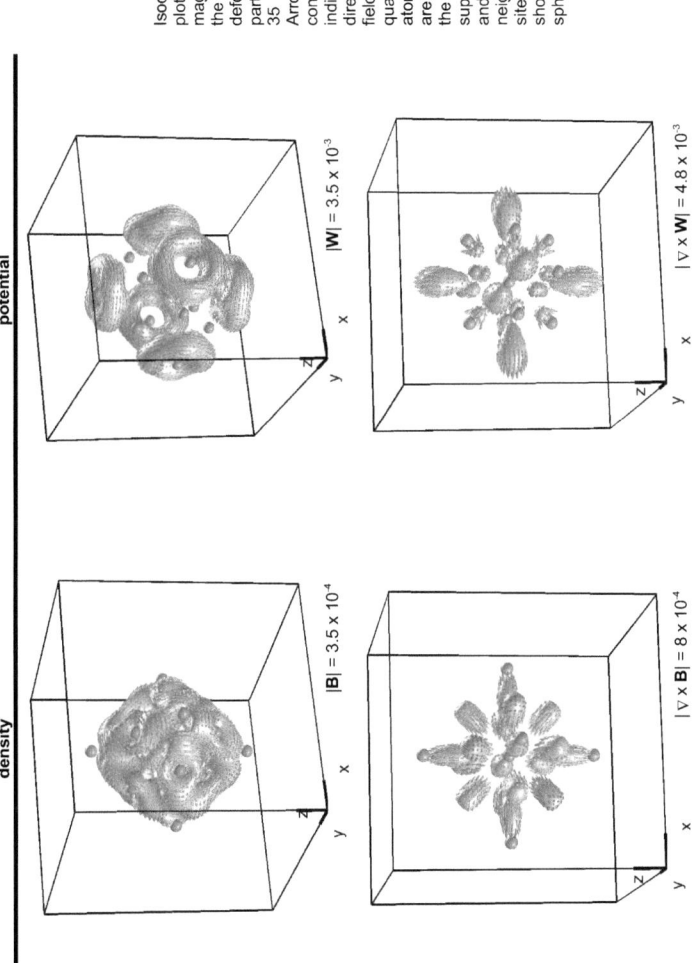

Isocontour plots of the magnitude of the deformation part of Li at 35 GPa. Arrows on the contours indicate the direction of the fields. The quantities, in atomic units, are plotted in the 3x3 cubic supercell; first and second neighbor Li sites are shown as small spheres.

168 APPENDIX A. ENATOM QUANTITIES

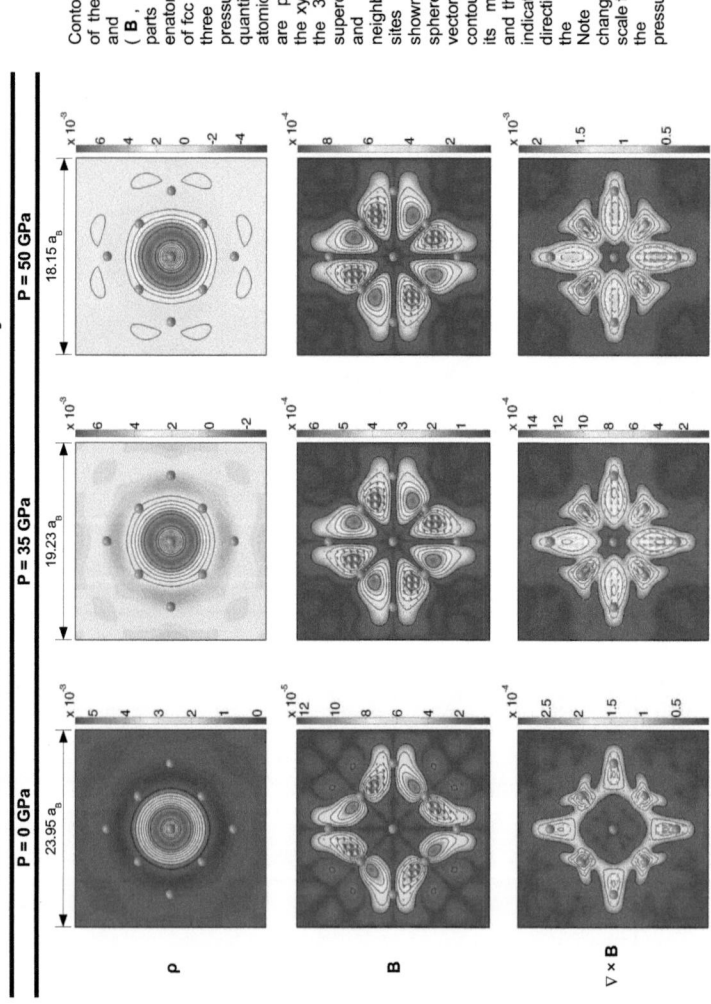

The Pressure Evolution of the Enatom *Density* in Lithium

Contour plots of the rigid (ρ) and non-rigid ($\mathbf{B}$, $\nabla \times \mathbf{B}$) parts of the enatom density of fcc lithium at three different pressures. The quantities, in atomic units, are plotted in the xy plane of the 3x3 cubic supercell; first and second neighbor Li sites are shown as small spheres. For a vector field the contour shows its magnitude, and the arrows indicate its direction within the plane. Note the change in scale for the different pressures.

169

The Pressure Evolution of the Enatom *Potential* in Lithium

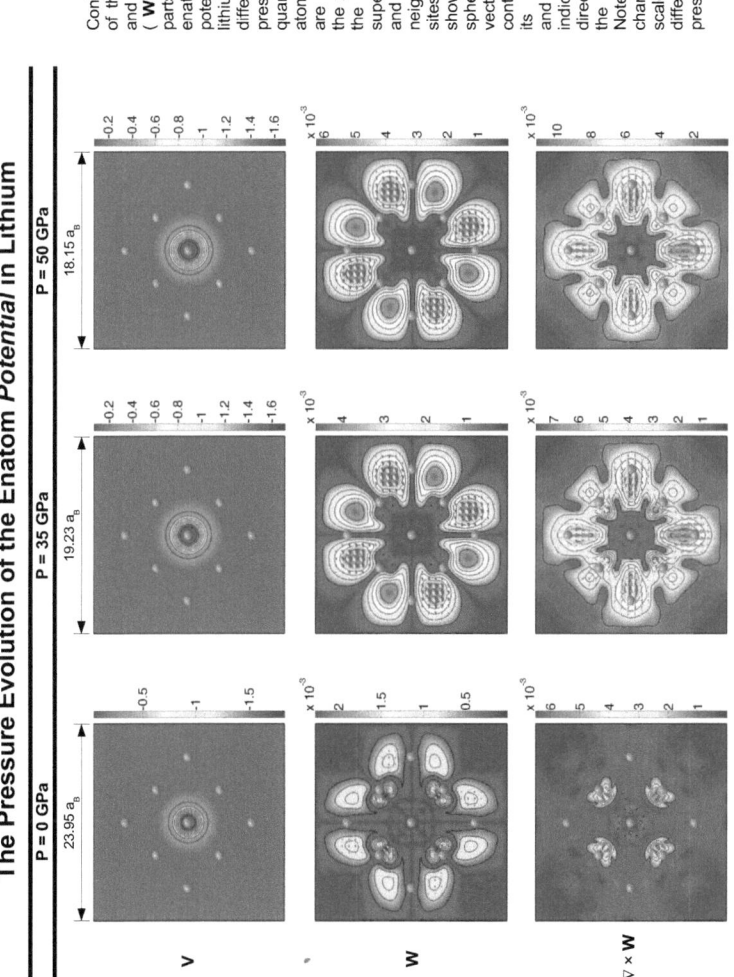

Contour plots of the rigid (V) and non-rigid (**W**, $\nabla \times \mathbf{W}$) parts of the enatom potential of fcc lithium at three different pressures. The quantities, in atomic units, are plotted in the xy plane of the 3x3 cubic supercell; first and second neighbor Li sites are shown as small spheres. For a vector field the contour shows its magnitude, and the arrows indicate its direction within the plane. Note the change in scale for the different pressures.

Bibliography

[1] Kroto, H. W., Heath, J. R., O'Brien, S. C., Curl, R. F., and Smalley, R. E. *Nature* **318**, 162 (1985).

[2] Iijima, S. *Nature* **354**, 56 (1991).

[3] Novoselov, K. S., Geim, A. K., Morozov, S. V., Jiang, D., Zhang, Y., Dubonos, S. V., Grigorieva, I. V., and Firsov, A. A. *Science* **306**, 666 (2004).

[4] Nagamatsu, J., Nakagawa, N., Muranaka, T., Zenitani, Y., and Akimitsu, J. *Nature* **410**, 63 (2001).

[5] Hanfland, M., Syassen, K., Christensen, N. E., and Novikov, D. L. *Nature* **408**, 174 (2000).

[6] Shimizu, K., Ishikawa, H., Takao, D., Yagi, T., and Amaya, K. *Nature* **419**, 597 (2002).

[7] Struzhkin, V. V., Eremets, M. I., Gan, W., Mao, H.-K., and Hemley, R. J. *Science* **298**, 1213 (2002).

[8] Deemyad, S. and Schilling, J. S. *Phys. Rev. Lett.* **91**, 167001 (2003).

[9] Eremets, M. I., Struzhkin, V. V., Mao, H., and Hemley, R. J. *Science* **293**, 272 (2001).

[10] Shirai, K., Masago, A., and Katayama-Yoshida, H. *phys. stat. sol. (b)* **244**, 303 (2007).

[11] Nelmes, R. J., Loveday, J. S., Allan, D. R., Besson, J. M., Hamel, G., Grima, P., and Hull, S. *Phys. Rev. B* **47**, 7668 (1993).

[12] Ma, Y., Prewitt, C. T., Zou, G., Mao, H., and Hemley, R. J. *Phys. Rev. B* **67**, 174116 (2003).

[13] Sanz, D. N., Loubeyre, P., and Mezouar, M. *Phys. Rev. Lett.* **89**, 245501 (2002).

[14] Matkovich, V. I., editor. *Boron and Refractory Borides*. Springer, Berlin, (1977).

[15] Pauling, L. *The Nature of the Chemical Bond*. Cornell University Press, Ithaca, (1960).

[16] Eberhardt, W. H., Crawford, B., and Libscomb, W. N. *J. Chem. Phys.* **22**, 989 (1954).

[17] Longuet-Higgins, H. C. *Q. Rev. Chem. Soc.* **11**, 121 (1957).

[18] Muetterties, E. L., editor. *The Chemsitry of Boron and its Compounds*. Wiley, New York, (1967).

[19] Zhai, H. J., Kiran, B., Li, J., and Wang, L. S. *Nat. Mater.* **2**, 827 (2003).

[20] Kiran, B., Bulusu, S., Zhai, H. J., Yoo, S., Zeng, X. C., and Wang, L. S. *Proc. Natl. Acad. Sci. U.S.A.* **102**, 961 (2005).

[21] Ciuparu, D., Klie, R. F., Zhu, Y., and Pfefferle, L. *J. Phys. Chem. B* **108**, 3967 (2004).

[22] Boustani, I. *Phys. Rev. B* **55**, 16426 (1997).

[23] Boustani, I. *Surf. Sci.* **370**, 355 (1997).

[24] Boustani, I. and Quandt, A. *Europhys. Lett.* **39**, 527 (1997).

[25] Gindulyte, A., Lipscomb, W. N., and Massa, L. *Inorg. Chem* **37**, 6544 (1998).

[26] Boustani, I. *J. Solid State Chem.* **133**, 182 (1997).

[27] Feynman, R. *Engineering and Science* Feb (1960). California Institute of Technology.

[28] Moore, G. E. *Electronics Magazine* **38** April (1965).

[29] Saito, R., Dresselhaus, G., and Dresselhaus, M. S. *Physical Properties of Carbon Nanotubes*. Imperial College Press, London, (1998).

[30] Iijima, S. and Ichihashi, T. *Nature* **363**, 603 (1993).

[31] Bethune, D. S., Klang, C. H., de Vries, M. S., Gorman, G., Savoy, R., Vazquez, J., and Beyers, R. *Nature* **363**, 605 (1993).

[32] Journet, C., Maser, W., Bernier, P., Loiseau, A., delaChapelle, M., Lefrant, S., Deniard, P., Lee, R., and Fischer, J. *Nature* **388**, 756 (1997).

[33] Szwacki, N. G., Sadrzadeh, A., and Yakobson, B. I. *Phys. Rev. Lett.* **98**, 166804 (2007).

[34] Szwacki, N. G. *Nanoscale Res. Lett.* **3**, 49 (2008).

[35] Donohue, J. *The Structure of the Elements*. Wiley, New York, (1974).

[36] Ekimov, E. A., Sidorov, V. A., Bauer, E. D., Mel'nik, N. N., Curro, N. J., Thompson, J. D., and Stishov, S. M. *Nature* **428**, 542 (2004).

[37] Boeri, L., Kortus, J., and Andersen, O. K. *Phys. Rev. Lett.* **93**, 237002 (2004).

[38] Lee, K.-W. and Pickett, W. E. *Phys. Rev. Lett.* **93**, 237003 (2004).

[39] Zhao, J. and Lu, J. P. *Phys. Rev. B* **66**, 092101 (2002).

[40] Calandra, M., Vast, N., and Mauri, F. *Phys. Rev. B* **69**, 224505 (2004).

[41] Mailhiot, C., Grant, J. B., and McMahan, A. K. *Phys. Rev. B* **42**, 9033 (1990).

[42] Papaconstantopoulos, D. A. and Mehl, M. J. *Phys. Rev. B* **65**, 172510 (2002).

[43] Bose, S. K., Kato, T., and Jepsen, O. *Phys. Rev. B* **72**, 184509 (2005).

[44] Häussermann, U., Simak, S. I., Ahuja, R., and Johansson, B. *Phys. Rev. Lett.* **90**, 065701 (2003).

[45] Segall, D. E. and Arias, T. A. *Phys. Rev. B* **67**, 064105 (2003).

[46] Ma, Y., Tse, J. S., Klug, D. D., and Ahuja, R. *Phys. Rev. B* **70**, 214107 (2004).

[47] Baroni, S., de Gironcoli, S., Dal Corso, A., and Giannozzi, P. *Rev. Mod. Phys.* **73**, 515 (2001).

[48] Savrasov, S. Y. *Phys. Rev. B* **54**, 16470 (1996).

[49] Savrasov, S. Y. and Savrasov, D. Y. *Phys. Rev. B* **54**, 16487 (1996).

[50] Ball, M. A. *J. Phys. C* **8**, 3328 (1975).

[51] Ball, M. A. *J. Phys. C* **10**, 4921 (1977).

[52] Lang, K. M., Mizel, A., Mortara, J., Hudson, E., Hone, J., Cohen, M. L., Zettl, A., , and Davis, J. C. *J. Low Temp. Phys.* **114**, 445 (1999).

[53] Gubser, D. U. and Webb, A. W. *Phys. Rev. Lett.* **35**, 104 (1975).

[54] Akahama, Y., Nishimura, M., Kinoshita, K., Kawamura, H., and Ohishi, Y. *Phys. Rev. Lett.* **96**, 045505 (2006).

[55] Profeta, G., Franchini, C., Lathiotakis, N. N., Floris, A., Sanna, A., Marques, M. A. L., Lüders, M., Massidda, S., Gross, E. K. U., and Continenza, A. *Phys. Rev. Lett.* **96**, 047003 (2006).

[56] Kasinathan, D., Kuneš, J., Lazicki, A., Rosner, H., Yoo, C. S., Scalettar, R. T., and Pickett, W. E. *Phys. Rev. Lett.* **96**, 047004 (2006).

[57] Kasinathan, D., Koepernik, K., Kuneš, J., Rosner, H., and Pickett, W. *Physica C* **460-462**, 133 (2007).

[58] Maheswari, S. U., Nagara, H., Kusakabe, K., and Suzuki, N. *J. Phys. Soc. Jpn.* **74**, 3227 (2005).

[59] Hohenberg, P. and Kohn, W. *Phys. Rev.* **136**, B864 (1964).

[60] Kohn, W. and Sham, L. J. *Phys. Rev.* **140**, A1133 (1965).

[61] von Barth, U. and Hedin, L. *J. Phys. C* **5**, 1629 (1972).

[62] Perdew, J. P. and Zunger, A. *Phys. Rev. B* **23**, 5048 (1981).

[63] Langreth, D. C. and Mehl, M. J. *Phys. Rev. B* **28**, 1809 (1983).

[64] Perdew, J. P. and Wang, Y. *Phys. Rev. B* **33**, 8800 (R) (1986).

[65] Perdew, J. P., Burke, K., and Ernzerhof, M. *Phys. Rev. Lett.* **77**, 3865 (1996).

[66] Anisimov, V. I., Zaanen, J., and Andersen, O. K. *Phys. Rev. B* **44**, 943 (1991).

[67] Anisimov, V. I., Solovyev, I. V., Korotin, M. A., Czyżyk, M. T., and Sawatzky, G. A. *Phys. Rev. B* **48**, 16929 (1993).

[68] Born, M. and Huang, K. *Dynamical Theory of Crystal Lattices*. Clarendon Press, Oxford, (1954).

[69] Hellmann, H. *Einführung in die Quantenchemie*. Deuticke, Leipzig, (1937).

[70] Feynman, R. P. *Phys. Rev.* **56**, 340 (1939).

[71] Pulay, P. *Mol. Phys.* **17**, 197 (1969).

[72] Mackintosh, A. K. and Andersen, O. K. In *Electrons at the Fermi Surface*, Springford, M., editor. Cambridge University Press, Cambridge (1975).

BIBLIOGRAPHY 175

[73] Gräfenstein, J. and Ziesche, P. *Phys. Rev. B* **53**, 7143 (1996).

[74] Press, W. H., Teukolsky, S. A., Vetterling, W. T., and Flannery, B. P. *Numerical Recipes*. Cambridge University Press, Cambridge, (1995).

[75] Ashcroft, N. W. and Mermin, N. D. *Solid State Physics*. Hartcourt, Orlando, (1976).

[76] Thomas, L. H. *Proc. Camp. Philos. Soc.* **23**, 542 (1927).

[77] Fermi, E. *Rend. Accad. Naz. Lincei* **6**, 602 (1927).

[78] Parr, R. G. and Yang, W. *Density Functional Theory of Atoms and Molecules*. Oxford University Press, Oxford, (1989).

[79] Eschrig, H. *The Fundamentals of Density Functional Theory*. B.G. Teubner Verlagsgesellschaft, Leipzig, (1996).

[80] Perdew, J. P. and Kurth, S. In *A Primer in Density Functional Theory*, Fiolhais, C., Nogueira, N., and Marques, M., editors. Springer-Verlag, Heidelberg (2003).

[81] Ceperley, D. M. and Alder, B. J. *Phys. Rev. Lett.* **45**, 566 (1980).

[82] Sternheimer, R. M. *Phys. Rev.* **96**, 951 (1954).

[83] Allen, P. B. *Phys. Rev. B* **6**, 2577 (1972).

[84] Midgal, A. B. *Zh. Eksp. Teor. Fiz.* **34**, 1438 (1958).

[85] Eliashberg, G. M. *Zh. Eksp. Teor. Fiz.* **38**, 966 (1960).

[86] Eliashberg, G. M. *Zh. Eksp. Teor. Fiz.* **39**, 1437 (1960).

[87] Bardeen, J., Cooper, L. N., and Schrieffer, J. R. *Phys. Rev.* **108**, 1175 (1957).

[88] Schrieffer, J. R. *Theory of Superconductivity*. Benjamin, New York, (1964).

[89] Allen, P. B. and Mitrović, B. In *Solid State Physics*, Ehrenreich, H., Seitz, F., and Turnball, D., editors, volume 37. Academic, New York (1982).

[90] McMillan, W. L. *Phys. Rev.* **167**, 331 (1968).

[91] Andersen, O. K. In *Computational Methods in Band Theory*, Markus, P. M., Janak, J. F., and Williams, A. R., editors. Plenum, New York (1971).

[92] Andersen, O. K. *Phys. Rev. B* **12**, 3060 (1975).

[93] Slater, J. C. *Phys. Rev.* **51**, 846 (1937).

[94] Korringa, J. *Physica* **13**, 392 (1947).

[95] Kohn, W. and Rostoker, N. *Phys. Rev.* **94**, 1111 (1954).

[96] Skriver, H. L. *The LMTO Method*. Springer, Berlin, (1984).

[97] Andersen, O. K. In *The Electronic Structure of Complex Systems,* Phariseau, P. and Temmerman, W. M., editors. Plenum, New York (1984).

[98] Andersen, O. K. and Jepsen, O. *Phys. Rev. Lett.* **53**, 2571 (1984).

[99] Andersen, O. K., Skriver, H. L., Nohl, H., and Johansson, B. *Pure Appl. Chem.* **52**, 93 (1979).

[100] Andersen, O. K., Jepsen, O., and Sob, M. In *Lecture Notes in Physics: Electronic Band Structure and Its Applications.,* Yussouff, M., editor. Springer-Verlag, Berlin (1987).

[101] Fermi, E. *Il Nuovo Cimento* **11**, 157 (1934).

[102] Phillips, J. C. and Kleinman, L. *Phys. Rev.* **116**, 287 (1959).

[103] Herring, C. *Phys. Rev.* **57**, 1169 (1940).

[104] Kleinman, L. and Bylander, D. M. *Phys. Rev. Lett.* **48**, 1425 (1982).

[105] Troullier, N. and Martins, J. L. *Phys. Rev. B* **43**, 1993 (1991).

[106] Vanderbilt, D. *Phys. Rev. B* **41**, 7892 (R) (1990).

[107] Pickett, W. E. *Comp. Phys. Rep.* **9**, 115 (1989).

[108] Singh, D. J. *Planewaves, Pseudopotentials and the LAPW Method*. Kluwer Academic Publishers, Dordrecht, (1994).

[109] Nogueira, F., Castro, A., and Marques, M. A. L. In *A Primer in Density Functional Theory,* Fiolhais, C., Nogueira, N., and Marques, M., editors. Springer-Verlag, Heidelberg (2003).

[110] Quandt, A. and Boustani, I. *ChemPhysChem* **6**, 2001 (2005).

[111] Exner, K. and von Ragué Schleyer, P. *Science* **290**, 1937 (2000).

[112] Wang, Z. X. and von Ragué Schleyer, P. *Science* **292**, 2465 (2001).

[113] Will, G. and Kiefer, B. *Z. Anorg. Allg. Chem.* **627**, 2100 (2001).

[114] Slack, G. A., Hejna, C. I., Garbauskas, M. F., and Kasper, J. S. *J. Solid State Chem.* **76**, 52 (1988).

[115] Hoard, J. L., Hughes, R. E., and Sands, D. E. *J. Amer. Chem. Soc.* **80**, 4507 (1958).

[116] Vlasse, M., Naslain, R., Kasper, J. S., and Ploog, K. *J. Solid State Chem.* **28**, 289 (1979).

[117] Hahn, T., editor. *International Tables for Crystallography*, volume A. Riedel, Boston, (1987).

[118] Muetterties, E. L. In *The Chemsitry of Boron and its Compounds*, Muetterties, E. L., editor. Wiley, New York (1967).

[119] Laubengayer, A. W., Hurd, D. T., Newkirk, A. E., and Hoard, J. L. *J. Am. Chem. Soc.* **65**, 1924 (1943).

[120] Bullett, D. W. *J. Phys. C* **15**, 415 (1982).

[121] Fujimori, M., Nakata, T., Nakayama, T., Nishibori, E., Kimura, K., Takata, M., and Sakata, M. *Phys. Rev. Lett.* **82**, 4452 (1999).

[122] Kimura, K. *Mat. Sci. Eng. B-Solid* **19**, 67 (1993).

[123] Weygand, D. *Ordre Local Icosaédrique dans des cristaux da base bore et modélisation de quasicristaux de bore.* PhD thesis, Institut National Polytechnique de Grenoble, (1994).

[124] Shirai, K., Masago, A., and Katayama-Yoshida, H. *phys. stat. sol. (b)* **241**, 3167 (2004).

[125] Suslick, K. http://www.scs.uiuc.edu/ chem315/handouts/ho_betaB.htm, (2001).

[126] Shirai, K. *Phys. Rev. B* **55**, 12235 (1997).

[127] Ploog, K., Schmidt, H., Amberger, E., Will, G., and Kossobutzki, K. H. *J. Less Common Met.* **29**, 161 (1972).

[128] Naslain, R. In *Boron and Refractory Borides*, Matkovich, V. I., editor. Springer, Berlin (1977).

[129] Longuet-Higgins, H. C. and de V. Roberts, M. *Proc. R. Soc. London., Series A* **230**, 110 (1955).

[130] Wikipedia. http://en.wikipedia.org.

[131] Rösler, H. J. *Lehrbuch der Mineralogie*. Deutscher Verlag für Grundstoffindustrie, Leipzig, (1988).

[132] Hoard, J. L. and Hughes, R. E. In *The Chemsitry of Boron and its Compounds*, Muetterties, E. L., editor. Wiley, New York (1967).

[133] Masago, A., Shirai, K., and Katayama-Yoshida, H. *Mol. Simul.* **30**, 935 (2004).

[134] Masago, A., Shirai, K., and Katayama-Yoshida, H. *Phys. Rev. B* **73**, 104102 (2006).

[135] van Setten, M. J., Uijttewaal, M. A., de Wijs, G. A., and de Groot, R. A. *J. Am. Chem. Soc.* **129**, 2458 (2007).

[136] Emin, D. *Phys. Today* **20**(1), 55 (1987).

[137] Nelmes, R. J., Loveday, J. S., Wilson, R. M., Marshall, W. G., Besson, J. M., Klotz, S., Hamel, G., Aselage, T. L., and Hull, S. *Phys. Rev. Lett.* **74**, 2268 (1995).

[138] Lazzari, R., Vast, N., Besson, J. M., Baroni, S., and Dal Corso, A. *Phys. Rev. Lett.* **83**, 3230 (1999).

[139] García, A. and Cohen, M. L. *Phys. Rev. B* **47**, 4215 (1993).

[140] He, J., Wu, E., Wang, H., Liu, R., and Tian, Y. *Phys. Rev. Lett.* **94**, 015504 (2005).

[141] Hanley, L. and Anderson, S. L. *J. Chem. Phys.* **89**, 2848 (1988).

[142] Hanley, L., Whitten, J. L., and Anderson, S. L. *J. Phys. Chem.* **92**, 5803 (1988).

[143] Boustani, I. *Int. J. Quantum Chem.* **52**, 1081 (1994).

[144] Boustani, I. *Chem. Phys. Lett.* **233**, 273 (1995).

[145] Boustani, I. *Chem. Phys. Lett.* **240**, 135 (1995).

[146] Boustani, I., Quandt, A., and Kramer, P. *Europhys. Lett.* **36**, 583 (1996).

[147] Boustani, I., Quandt, A., and Rubio, A. *J. Solid State Chem.* **154**, 269 (2000).

[148] Lipscomb, W. N. and Massa, L. *Inorg. Chem.* **31**, 2297 (1992).

[149] Lipscomb, W. N. and Massa, L. *Inorg. Chem.* **33**, 5155 (1994).

[150] Derecskei-Kovacs, A., Dunlap, B. I., Lipscomb, W. N., Lowrey, A., Marynick, D. S., and Massa, L. *Inorg. Chem.* **33**, 5617 (1994).

[151] Gindulyte, A., Krishnamachari, N., Lipscomb, W. N., and Massa, L. *Inorg. Chem* **37**, 6546 (1998).

[152] Jemmis, E. D. and Jayasree, E. *Acc. Chem. Res.* **36**, 816 (2003).

[153] Tang, A. C., Li, Q. S., Liu, C. W., and Li, J. *Chem. Phys. Lett.* **201**, 465 (1993).

[154] Aihara, J., Kanno, H., and Ishida, T. *J. Am. Chem. Soc* **127**, 13324 (2005).

[155] Cao, L., Zhang, Z., Sun, L., Gao, C., He, M., Wang, Y., Li, Y., Zhang, X., Li, G., Zhang, J., and Wang, W. *Adv. Mater.* **13**, 1701 (2001).

[156] Wu, Y., Messer, B., and Yang, P. *Adv. Mater.* **13**, 1487–1489 (2001).

[157] Otten, C. J., Lourie, O. R., Yu, M., Cowley, J. M., Dyer, M. J., Ruoff, R. S., and Buhro, W. E. *J. Am. Chem. Soc.* **124**, 4564 (2002).

[158] Wang, Z., Shimizu, Y., Sasaki, T., Kawaguchi, K., Kimura, K., and Koshizaki, N. *Chem. Phys. Lett.* **368**, 663 (2003).

[159] Xu, T. T., Zheng, J.-G., Wu, N., Nicholls, A. W., Roth, J. R., Dikin, D. A., and Ruoff, R. S. *Nano Lett.* **4**, 963 (2004).

[160] Fowler, J. E. and Ugalde, J. M. *J. Phys. Chem. A* **104**, 397 (2000).

[161] Ritter, S. K. *Chem. Eng. News* **82**, 28 (2004).

[162] Kresse, G. and Furthmüller, J. *Comput. Mater. Sci.* **6**, 15 (1996).

[163] Kresse, G. and Furthmüller, J. *Phys. Rev. B* **54**, 11169 (1996).

[164] Payne, M. C., Teter, M. P., Allan, D. C., Arias, T. A., and Joannopoulos, J. D. *Rev. Mod. Phys.* **64**, 1045 (1992).

[165] Perdew, J. P. and Wang, Y. *Phys. Rev. B* **45**, 13244 (1992).

[166] Kresse, G. and Hafner, J. *J. Phys.: Condens. Matter* **6**, 8245 (1994).

[167] Methfessel, M. and Paxton, A. T. *Phys. Rev. B* **40**, 3616 (1989).

[168] Becke, A. D. and Edgecombe, K. E. *J. Chem. Phys.* **92**, 5397 (1990).

[169] Jepsen, O. and Andersen, O. K. *Solid State Commun.* **9**, 1763 (1971).

[170] Perdew, J. P., Burke, K., and Wang, Y. *Phys. Rev. B* **54**, 16533 (1996).

[171] Kokalj, A. *Comput. Mater. Sci.* **28**, 155 (2003).

[172] Stokes, H. T., Hatch, D. M., and Campbell, B. J. http://stokes.byu.edu/isotropy.html, (2007).

[173] Evans, M. H., Joannopoulos, J. D., and Pantelides, S. T. *Phys. Rev. B* **72**, 45434 (2005).

[174] Cabria, I., López, M. J., and Alonso, J. A. *Nanotechnology* **17**, 778 (2006).

[175] Lau, K. C., Pati, R., and ans A. C. Pineda, R. P. *Chem. Phys. Lett.* **418**, 549 (2006).

[176] Boustani, I. and Quandt, A. *Comp. Mater. Sci.* **11**, 132 (1998).

[177] Boustani, I., Quandt, A., Hernandez, E., and Rubio, A. *J. Chem. Phys.* **110**, 3176 (1999).

[178] Kunstmann, J. and Quandt, A. *Chem. Phys. Lett.* **402**, 21 (2005).

[179] Leys, F. E., Amovilli, C., and March, N. H. *J. Chem. Inf. Comput. Sci.* **44**, 122 (2004).

[180] Damnjanović, M., Milošević, I., Vuković, T., and Sredanović, R. *Phys. Rev. B* **60**, 2728 (1999).

[181] Milošević, I. and Damnjanović, M. *J. Phys.: Condens. Matter* **18**, 8139 (2006).

[182] Bagci, V. M. K., Gülseren, O., Yildirim, T., Gedik, Z., and Ciraci, S. *Phys. Rev. B* **66**, 45409 (2002).

[183] Miyamoto, Y., Rubio, A., Louie, S. G., and Cohen, M. L. *Phys. Rev. B* **60**, 13885 (1999).

[184] White, C. T., Robertson, D. H., and Mintmire, J. W. *Phys. Rev. B* **47**, 5485 (1993).

[185] Lin-Chung, P. J. and Rajagopal, A. K. *J. Phys.: Condens. Matter* **6**, 3679 (1994).

[186] Quandt, A., Liu, A. Y., and Boustani, I. *Phys. Rev. B* **64**, 125422 (2001).

[187] Chopra, N. G., Benedict, L. X., Crespi, V. H., Cohen, M. L., Louie, S. G., and Zettl, A. *Nature* **337**, 135 (1995).

[188] Elliott, J. A., Sandler, J. K. W., Windle, A. H., Young, R. J., and Shaffer, M. S. P. *Phys. Rev. Lett.* **92**, 095501 (2004).

[189] Gerlich, D. and Slack, G. A. *J. Mater. Sci. Lett.* **4**, 639 (1985).

[190] Teter, D. M., Gibbs, G. V., Boisen, M. B., Allan, D. C., and Teter, M. P. *Phys. Rev. B* **52**, 8064 (1995).

[191] Murnaghan, F. D. *Proc. Natl. Acad. Sci.* **30**, 244 (1944).

[192] Kong, Y., Dolgov, O. V., Jepsen, O., and Andersen, O. K. *Phys. Rev. B* **64**, 020501(R) (2001).

[193] Dolgov, O. V., Andersen, O. K., and Mazin, I. I. arXiv:0710.0661v1, (2008).

[194] Miller, S. C. and Love, W. F. *Tables of Irreducible Representations of Space Groups and Co-representations of Magnetic Space Groups*. Pruett Pr., Boulder, (1967).

[195] Dresselhaus, M. S., Dresselhaus, G., and Eklund, P. *Science of Fullerenes and Carbon Nanotubes*. Academic Press, San Diego, (1996).

[196] Tang, H. and Ismail-Beigi, S. *Phys. Rev. Lett.* **99**, 115501 (2007).

[197] Yang, X., Ding, Y., and Ni, J. *Phys. Rev. B* **77**, 041402(R) (2008).

[198] Halford, B. *Chemical & Engineering News* **83**(35), 30 (2005).

[199] Marquit, M. *physorg.com* 27. Sept (2007).

[200] Sedgemore, F. *Nanomaterials News* **3**(17), 2 (2007).

[201] Miller, J. *Phys. Today* **60**(11), 20 (2007).

[202] Battersby, S. *New Scientist Technology* 4. Jan (2008).

[203] Zeng, W. *Nature China* 16. Jan (2008).

[204] Tenne, R., Margulis, L., Genut, M., and Hodes, G. *Nature* **360**, 444 (1992).

[205] Rubio, A., Corkill, J. L., and Cohen, M. L. *Phys. Rev. B* **49**, 5081 (1994).

[206] Chopra, N. G., Luyken, R. J., Cherrey, K., Crespi, V. H., Cohen, M. L., Louie, S. G., and Zettl, A. *Science* **269**, 966 (1995).

[207] Tang, Z. K., Sun, H. D., Wang, J., Chen, J., and Li, G. *Appl. Phys. Lett.* **73**, 2287 (1998).

[208] Robertson, D. H., Brenner, D. W., and Mintmire, J. W. *Phys. Rev. B* **45**, 12592 (1992).

[209] Hernández, E., Goze, C., Bernier, P., and Rubio, A. *Phys. Rev. Lett.* **80**, 4502 (1998).

[210] Seifert, G., Terrones, H., Terrones, M., Jungnickel, G., and Frauenheim, T. *Phys. Rev. Lett.* **85**, 146 (2000).

[211] Lipscomb, W. N. *Acc. Chem. Res.* **6**, 257 (1973).

[212] Kunstmann, J. and Quandt, A. *J. Chem. Phys.* **121**, 10680 (2004).

[213] Bailey, C. *The Greek Atomists and Epicurus*. Clarendon, Oxford, (1928).

[214] Ziman, J. M. *Adv. Phys.* **13**, 89 (1964).

[215] Dagens, L. *J. Phys. C* **5**, 2333 (1972).

[216] Streitenberger, P. *Phys. Status Solidi B* **116**, 179 (1983).

[217] Pickett, W. E. *J. Phys. C* **12**, 1491 (1979).

[218] Khan, F. S. and Allen, P. B. *Phys. Rev. B* **29**, 3341 (1984).

[219] Moussa, J. E. and Cohen, M. L. *Phys. Rev. B* **74**, 094520 (2006).

[220] Sham, L. J. *Phys. Rev.* **188**, 1431 (1969).

[221] Martin, R. M. *Phys. Rev. B* **5**, 1607 (1972).

[222] Falter, C., Ludwig, W., Maradudin, A. A., Selmke, M., and Zierau, W. *Phys. Rev. B* **32**, 6510 (1985).

[223] Falter, C., Rakel, H., Klenner, M., and Ludwig, W. *Phys. Rev. B* **40**, 7727 (1989).

[224] Falter, C., Selmke, M., Ludwig, W., and Kunc, K. *Phys. Rev. B* **32**, 6518 (1985).

[225] Ball, M. A. and Srivastava, G. P. *J. Phys.: Condens. Matter* **5**, 2511 (1993).

[226] Ball, M. A. and Srivastava, G. P. *J. Phys.: Condens. Matter* **4**, 1947 (1992).

[227] Hamlin, J. J., Tissen, V. G., and Schilling, J. S. *Phys. Rev. B* **73**, 094522 (2006).

[228] Yabuuchi, T., Matsuoka, T., Nakamoto, Y., and Shimizu, K. *J. Phys. Soc. Jpn.* **75**, 083703 (2006).

[229] Shi, L. and Papaconstantopoulos, D. A. *Phys. Rev. B* **73**, 184516 (2006).

[230] Christensen, N. E. and Novikov, D. L. *Phys. Rev. B* **73**, 224508 (2006).

[231] Baroni, S., Corso, A. D., de Gironcoli, S., Giannozzi, P., Cavazzoni, C., Ballabio, G., Scandolo, S., Chiarotti, G., Focher, P., and *et al.*, A. P. http://www.pwscf.org.

[232] Monkhorst, H. J. and Pack, J. D. *Phys. Rev. B* **13**, 5188 (1976).

[233] Marzari, N., Vanderbilt, D., Vita, A. D., and Payne, M. C. *Phys. Rev. Lett.* **82**, 3296 (1999).

[234] Syassen, K. and Holzapfel, W. B. *J. Appl. Phys.* **49**, 4427 (1978).

[235] Gaspari, G. D. and Gyorffy, B. L. *Phys. Rev. Lett.* **28**, 801 (1972).

[236] von der Lage, F. C. and Bethe, H. A. *Phys. Rev.* **71**, 612 (1947).

[237] Lehmann, K. K. and Callegan, C. *J. Chem. Phys.* **117**, 1595 (2002).

[238] Lau, K. C. and Pandey, R. *J. Phys. Chem. C* **111**, 2906 (2007).

[239] Kunstmann, J. and Quandt, A. *Phys. Rev. B* **74**, 035413 (2006).

[240] Kunstmann, J., Quandt, A., and Boustani, I. *Nanotechnology* **18**, 155703 (2007).

[241] Kunstmann, J., Boeri, L., and Pickett, W. E. *Phys. Rev. B* **75**, 075107 (2007).

Acknowledgments

Now this is the end. There is nothing else left to say but "Thanks" to everyone who supported me during my time in Stuttgart.

- First of all I want to express my gratitude to my supervisor Prof. Dr. Ole Krogh Andersen for giving me the chance to work in his group. He allowed me to go my own way, which I appreciated very much. And whenever I wanted to go to a conference or invite people to Stuttgart, he made it possible. Thanks a lot!

- I thank Prof. Dr. Alejandro Muramatsu for undertaking the task of being the coreferee of this thesis.

- Many thanks also go to Prof. Dr. Warren E. Pickett from the U. C. Davis. During his sabbatical in Stuttgart, he taught my colleague Lilia Boeri and me the "magic" of the generalized pseudoatom of M. A. Ball. Later we had lots of fun making colorful figures and inventing the name "enatom". Among us: I think this was a great marketing concept.

- Thanks to Prof. Dr. Jens Kortus from the TU Freiberg for drawing my interest to high–pressure boron and helpful and motivating discussions.

- This thesis would be nothing without the help, patience, and cleverness of my dear friend and colleague Lilia Boeri, who somehow became my "unofficial" day–to–day supervisor. So many problems seemed unsolvable to me, but after Lilia had a look at them, there would always be a way out. Lilia, you are amazing!

- I want to thank the usual boron suspects, i.e., my friends and colleagues Dr. Alexander Quandt from the University of Greifswald and Dr. Ihsan Boustani from the University of Wuppertal for their constant support and collaboration. The truth is that we are all addicted to boron!

- I am also very happy that I could do my PhD in the Andersen department. You'll rarely find a group with such a nice and friendly atmosphere in good old Germany anymore. Special thanks go to our always supportive secretary

Claudia Hagemann and to Dr. Ove Jepsen for helping me so many times with his huge expertise in electronic structure calculations. Ove, I should admit that I started the rebellion against Danish butter cookies. But I think you already gave De Beukelaer "Prinzelrolle" a chance.

- I also want to thank the superb computer service group. Whenever there was a problem with the computers they would solve it promptly. I can not express how much I appreciated that. People, you are great!

- Being a fellow of the International Max Planck Research School for Advanced Materials was a great experience. I loved the international atmosphere, the kindness of all my fellow students, and the time we spend together. Many thanks to the coordinator Dr. Hans–Georg Libuda.

- I want to thank God, for being there at all the times. After all it's you who makes everything possible.

- As this thesis started with the dedication to my family, it should also end with it: *Zum Abluss möchte ich meiner ganzen Familie für ihre Unterstützung danken. Ganz besonders ihrem jüngsten Mitglied, meinem Mann Martin. Danke für's Warten.*

VDM Verlagsservicegesellschaft mbH

Die VDM Verlagsservicegesellschaft sucht für wissenschaftliche Verlage abgeschlossene und herausragende

Dissertationen, Habilitationen, Diplomarbeiten, Master Theses, Magisterarbeiten usw.

für die kostenlose Publikation als Fachbuch.

Sie verfügen über eine Arbeit, die hohen inhaltlichen und formalen Ansprüchen genügt, und haben Interesse an einer honorarvergüteten Publikation?

Dann senden Sie bitte erste Informationen über sich und Ihre Arbeit per Email an *info@vdm-vsg.de*.

Sie erhalten kurzfristig unser Feedback!

VDM Verlagsservicegesellschaft mbH
Dudweiler Landstr. 99
D - 66123 Saarbrücken

Telefon +49 681 3720 174
Fax +49 681 3720 1749

www.vdm-vsg.de

Die VDM Verlagsservicegesellschaft mbH vertritt

Printed by Books on Demand GmbH, Norderstedt / Germany